C000132761

CHASING RAINBOWS

CHASING RAINBOWS

HOW THE GREEN AGENDA
DEFEATS ITS AIMS

Tim Worstall

STACEY
INTERNATIONAL

Chasing Rainbows

STACEY INTERNATIONAL
128 Kensington Church Street
London W8 4BH
Tel: +44 (0)20 7221 7166; Fax: +44 (0)20 7792 9288
Email: info@stacey-international.co.uk
www.stacey-international.co.uk

ISBN: 978 1 906768 44 7

CIP Data: A catalogue record for this book is available from the British Library

1 3 5 7 9 0 8 6 4 2

Printed in Turkey

CONTENTS

1 Introduction

Green technologies also seem to provide plenty of jobs. Exploiting renewables now employs 2.3 million people worldwide[1], more than the entire oil and gas industries, even though they contribute a small fraction of the amount of energy. They provide several times as much work per dollar invested than fossil fuels, with other green measures like recycling and saving energy proving even more job-intensive.

Geoffrey Lean, *Daily Telegraph*[1]

That's the sort of drivel that produces conniption fits in economists. For as every purveyor of the dismal science will point out, jobs are a cost of a plan, not a benefit. We are thus attempting to make everything that we do less job intensive, not more so. To welcome something as requiring more effort to achieve than another method of getting to the same end is nonsensical. To argue that using the labour of 2.3 million people to produce energy is a better idea than using the labour of 1 million, or three people, or 2.1 million or any other number lower than 2.3 million, is simply to be ignorant of the basic defining desire of all human beings: to get the most we can for the least effort.

This is only one of the gross misunderstandings of the real world that the various greens, Greens and environmentalists fall prey to. This book is about many of those misunderstandings: an attempt to point out where the courses of action proposed for us by those who would make the world better are diametrically opposed to the actual courses of action which would do so. A look at where the things we are told we should be doing are completely different from the things we should actually be doing in order to get to the desired goal. That goal, one I assume we all share, being a cleaner, richer, better in short, world for us and our descendants.

This first error is entirely common, this idea that creating more jobs in managing to reach some goal or other is a good idea. It's also entirely wrong.

We've known that it is wrong at least since Frederic Bastiat published "The Broken Window"[2] back in 1850.... 160 years would, you'd think, be long enough for an idea to percolate through the collective unconscious but apparently not. Bastiat's point was that we can all see the extra and new economic activity that goes on when a window gets broken. The glazier has to be paid for the new one to be put in, the glazier has more money which he spends on, well, being French, some combination of wine, women and *La Vie en Rose* perhaps. Everyone applauds this growth in the national production and regrets are there none.

Except Bastiat urges us, in one of the great insights which economists continually try to remind both themselves and the rest of us of, to look at not just what is visible, but what is invisible here? What would have happened if the window had not been broken? The glazier would have been out his 6 francs it is true (and that's 6 old francs, roughly spit now, which shows that economists haven't been all that good on the preventing inflation front over the past 160 years). But the window owner would have still had his 6 francs and he could have had his own escapade on the cash. We would have not just the experience with the drug addled singer, the women and the wine but we would still have a window doing its job of keeping out the caterwauling from La V'nR. We can see that the world is richer not because work has been created for the glazier to do but rather richer by the fact that the glazier has needed to do nothing.

This is closely related to the economists' concept of "opportunity cost" and opportunity cost is one of the basic two things you need to know about the subject. As Glen Whitman[3] has pointed out, you can play the "Two Things"[4] game with an expert in any area. Just what are the two things which encapsulate the entire subject? In boxing they are: 1) Hit and 2) Don't get hit – which seems to sum up the barbarity elegantly. Civil engineering tells us that water plus dirt is mud and that there's always mud.

The two things about economics are that

1.   Incentives matter
2.   There's no free lunch

Another way of putting this idea that there's no free lunch (or as Robert Heinlein put it, TANSTAAFL, There Ain't No Such Thing As A Free Lunch) is that there are always opportunity costs. We cannot use the same time to do two different things[5], we cannot use the same resources to do two entirely different things. Once we've committed ourselves to go skiing that afternoon we cannot also go surfing: one of the prices of our going skiing is that we've given up going surfing. Once we've decided to use the labour of 2.3 million people to produce our renewable energy we cannot use that same labour to do something else.

This leads us to the true price of something. It's not the money we hand over, nor the time we've spent earning that money, that makes a pint of beer cost £4 or 40 minutes of minimum wage labour. The true price of something is what we give up to get it. It might be that £4 or 40 minutes, but it's also that Big Mac that we didn't buy instead of that pint of foaming that we could have had from that 40 minutes of labour.

Which is where we can begin to see why cheering the labour intensity of a particular technology, like renewable energy, is such a silly idea. We're giving up all of those other things that could have been created with that labour in order to get our energy from this technology. All those other things that we're not getting are therefore part of the cost of renewables: the jobs created are a cost of renewables, not a benefit.

Sadly, this nonsense is not unusual in the greener parts of the political spectrum. I once heard Caroline Lucas (Leader of the Green Party) insist on the *Today Programme* that, as Geoffrey Lean proposes at the top of this chapter, renewables are a good idea precisely because they are more labour intensive than other forms of energy generation.

Al Gore has had a whole campaign going called "We Can Solve It" (apparently Bob the Builder wasn't as popular over there as he was here for surely no one would try to echo his catchphrase?), the major selling point of which seems to be that[6]:

> Thousands of new companies, millions of new jobs, and billions in revenue generated by solutions to the climate crisis – this is the clean energy economy we can adopt with today's technologies, resources, know-how, and leadership from our elected officials.

Leave aside the ghastliness of our elected officials trying to actually run or create anything and note that he's suggesting that thousands of companies, millions of people and billions of dollars should be diverted into a less efficient form of generating energy, taking those thousands, millions and billions away from attempting to satisfy other human desires. To be fair though it would have been Bill running after the French singer. At least Al is a retired politician which, as I write, isn't true of Alistair Darling (although, as I write, it looks increasingly likely that he soon will be tending his dahlias) who said in his 2008 Budget Speech[7]:

> There are huge opportunities here too for business, and
> there could be over a million jobs in our environmental
> industries within the next two decades.

Again, that's a million people not doing something else; these
jobs are a cost to us, not a benefit. But then perhaps employing
a provincial solicitor as Chancellor of the Exchequer wasn't
all that good an idea in the first place: it's not as if he's had
much training at either the theory or the hard stuff like sums,
is it?

But how about Sir Jonathan Porritt Bt.? (Quite why this
Old Etonian's Baronetcy isn't more widely known I'm not sure.
Clearly it can't be because we might bridle at some scion of the
landed gentry telling us we should all be peasants again?)[8]:

> Wouldn't it be great, just once, to hear a senior Labour
> Politician (other than Ken) enthusing in similar terms
> about the hundreds of thousands of real jobs that would
> be created were we ever to get serious about energy
> efficiency?

There being hundreds of thousands of things not being done as
a result of everyone being off dealing with energy efficiency of
course. And there's no excuse for Porritt not getting it: he was,
at the time he said this, head of the Sustainable Development
Commission. You know, someone with actual power rather than
a mere politician, someone who was supposed to be advising
government on what actually to do about things rather than just
grin and posture for the electorate.

As our final example of the firm grasp of the ordure-stained
stick by the mucky end, may I present the new economics
foundation (they believe that a major ecological crisis is
approaching from the overuse of Capital Letters[9]) or nef. As
George Monbiot reports, they say that[10]:

> The New Economics Foundation has calculated that
> every £50,000 spent in small local shops creates one
> job. You must spend £250,000 in superstores for the
> same result.

The Guardian clearly is not so worried about the disturbing shortage of upper case typefaces and the nef here is not being factually incorrect. Their numbers are about right. It's just that they go on to insist that this means we should all shop at small shops and never darken the door of some ghastly chain. Despite the fact that our doing so would mean that for each £250 large we spend four more people have to be chained behind a store counter dealing with that most delightful of creatures, the Great British Public, for the entirety of their working lives. Rather than going and doing something more enjoyable like, say, being live models for voodoo dolls or something more useful like attempting to find the cure for cancer.

We shall meet the nef again, later in this volume, but the above alone might be enough for you to understand why Giles Wilkes[11] insists that the acronym really stands for Not Economics Frankly.

Just in case there are any still confused, perhaps Caroline or Jonathan have picked up this book by mistake or something, allow me to explain quite why directing people into more labour intensive jobs is, in and of itself, a bad thing (you at the nef can move your lips along as the rest of us read as well). We've got the theory above from a 19th century Frenchman but perhaps it should be put into terms they might actually understand.

Imagine an economy with 100 people in it. Now imagine three different levels of technology, three different levels of labour productivity.

Ah, no, let us take a little digression here about models. Yes, I'm aware, as anyone and everyone who ever uses a model is, that this model does not include the mess of the real world.

That's rather the point. We want to get rid of that mess for a bit and look just at one or two salient points that we want to discuss, emphasize. This is not meant to be a full and detailed explanation of all and everything, for that would be the reality that is so messy and which we all have such a hard time understanding. It's a thought experiment, nothing more, just an attempt to look at one or two factors. So don't start complaining that it doesn't include the dignity of labour, or the patriarchy of capitalism, the calories of oil we now use to grow calories of food or the way that hemp really could solve our problems. I know it doesn't include any of these, that's the point. *Capisce?*

Our three different scenarios, different technological situations:

In the first of our 100 people some 80 of them need to be working on the land to provide the food for the 100 to eat. The technology we use for farming isn't very good and a great deal of human labour has to be added to that technology (say, mostly bits of iron and wood strapped to large animals plus the mess that large animals tend to leave in fields) to produce the food. This leaves 20 people to provide all the manufactures and services that all 100 get to consume.

At the second level of technology only 20 people are needed to grow the food for the 100. 50 are off manufacturing things and 30 are providing services. Part of that manufacturing is of course of new and interesting machines to hoe turnips but there are also all sorts of other lovely things: plates that poor people can afford for example. Similarly services have expanded as there are more people available to provide them: it's now possible to get someone to wipe your forehead with a damp cloth as you twist and turn in that wonderful innovation, a hospital bed.

Our third scene, our third set of technologies, leaves us with only two people required to be standing in a muddy field to feed everyone. Productivity of labour in agriculture has risen very strongly. We also need only, just imagine, 18 people to produce all of the manufactured things we desire. This leaves 80 people to provide services. Including, just imagine, medical treatment that might actually cure you in that by now rather old hospital bed.

This is, with a little bit of ignoring things like imports and exports (it's a model, see?), roughly the economic history of the UK. Back in 1700 was our first case, the 1920s or so were the second and the third is now. If we translate the fixed number of people into percentages it's even more like our economic history. We started out needing the vast majority of the population to be working on the farms in order to produce food that would keep all alive. This meant there was very little labour that could be used to do other things, like manufacture things that could be dropped on people's feet or provide services. As time went on agriculture became more efficient as a user of labour and so there were more people able to go and work in factories making things rather than growing things. This past century has been the same with manufacturing. Productivity has been rising faster than our desire for more manufactured things and this has freed up labour to provide services.

We should also note here that while we've only got 2% or less of the population on the land we're growing a great deal more food than we were 300 years ago or even 90 years ago. It's also true that, whatever they say about the "evisceration" of manufacturing industry under various baby-eating Tories, manufacturing output by value is higher than it was in 1960, more than double what it was at the end of WWII and all while using fewer people to produce it. This means that we have, through this rise in labour productivity, more food and more manufactures and more services to enjoy: we're richer. Having more stuff is richer isn't it? Yes, we even have more of all these things after you take account of the rising population as well.

The rising productivity of labour over the centuries is therefore what has made us all richer. This is true even if you think that "more stuff" isn't what makes us richer, an idea that many of our green friends would support. We've, collectively, got to labour less to get that modicum of food, shelter and the essential bad silver jewelry/joss stick mix required to keep us going and have ever more leisure time with which to sing "Kumbaya". This might not be your or my idea of a joyous life, we might prefer that "more stuff" but as good little liberals we're not going to insist that others live as we desire to.

We can now see the problem with this mantra of "creating jobs". Using 2.3 million people to produce our non carbon energy requirements means that we're taking 2.3 million people away from doing other things which we might desire they be doing. This isn't, as alluded to above, simply a matter of those people not producing cars, or wide screen TVs. Even if those are the excessive consumerism that is oft decried, those 2.3 million people are also being taken away from, say, protesting against excessive consumerism, that vital part of a truly Green economy.

They are also being taken away from anything and everything else they could be doing: digging allotments, wiping the bottoms of babies, tending the elderly or simply sipping a quiet pint. The things we lose by allocating labour in this deliberately inefficient manner are thus costs of our non carbon energy system: undug allotments, unwiped babies, untended grannies and unsupped pints.

Now it is entirely possible for me to overdo this point. For of course we do actually need to do something: not just something about climate change or energy systems, but we do need to actually make a decision about what labour we're going to use to do what where. It's possible that using the labour of these people to produce energy without (or at least with low, for there are no "non carbon" energy systems) carbon emissions is what we really should be doing. But to do this we need a decision-making system, some method of weighing up the pros and cons of our various alternatives so that we can see which is the best one.

We have such a system: it's called a cost benefit analysis. We look at all of the costs of our potential course of action and then we look at all of the benefits of it. If the benefits are larger than the costs then it's a good thing to go and do. If the costs are higher than the benefits then it ain't. So we shouldn't go and do it. At this point we get into a slightly contentious area for economists do such analyses by reducing everything to money. This shocks some of the more maiden auntish of the population, that we would talk about such important things as Life On Planet Earth in terms of grubby and even filthy lucre. But this isn't what is actually being done. We're simply reducing values to something that can be compared, not trying to say that the polar bears are worth such and such a sum.

I once wrote on this subject for *The Guardian*'s "Comment is Free" site and up popped, in the comments, a self-proclaimed graduate degree holder in Philosophy claiming that the whole idea of cost benefit analyses was rubbish because we were trying to compare values which were not commensurable. Normally you expect philosophers (they are supposed to be trained in logic after all) to have rather better arguments than this. For what commensurable means is capable of being compared. So the statement is that you can't compare these costs and benefits because you can't compare these costs and benefits. Not really a huge advance in our ability to determine the correct course for the world, then.

However, what is meant isn't quite so silly. The economists are attempting to put a value on, if we're going to talk about climate change, Bangladesh sinking below the waves in 2100. The polar bears dying out by 2080. People dying tomorrow because they don't have vaccinations because we can't use fossil fuels to power the refrigerators needed to keep them. It does seem barking to assume that we can accurately place a value on a country, a species or a life. Which is why economists don't even attempt to do so.

What actually happens is that when a cost benefit analysis says that this marshland is worth £squiddelypop as a nesting and feeding area for wading birds the statement is not that you can take the marsh to the bank and in return get the cash for £squiddelypop divided by 4 pints of beer. Rather, it's that by their actions, on average, humans behave as if the value of this marshland as a nesting and feeding area for wading birds is worth that multiple of the value they put on that many pints of beer. We are not claiming that anything at all has an absolute value: we are claiming that human beings seem to put this value on it as compared to this other value they put on that over there. The only reason that we put these values into pounds and pence is so that we can do sums.

Another way of putting this is that we are using commensurable values because all of the values we're using are the (perhaps arbitrary, certainly subjective) values that human beings put on these things.

Now we have our method, this cost benefit analysis. We have the values which we can slot into our sums. We must now work out which is a cost and which is a benefit: sums notoriously don't work out if you get that bit wrong. And this is where the mantra of creating good green jobs goes so wrong (and the claim that high paid jobs are better than low paid takes us even further into the realms of nonsense: if we're paying more to get the job done this is a greater expense for us). For all the rhetoric is that the creation of these jobs is a benefit of our scheme. Which, as I hope the above has been able to convince you, isn't correct at all. All of these jobs created are a cost to us of this scheme. For by assigning labour to our pet project we've not only got to pay all of these labourers we also lose all of the other things that could have been produced for us by this same labour.

It's possible that the cost benefit analysis will none the less come out with it all being a good idea. If the benefit to us is of the survival of the human species then the cost of 2.3 million people, of what those 2.3 million could otherwise be doing, seems quite cheap in fact. If the benefit is that we've delayed a 0.1 degree Celsius temperature rise by 7 years then perhaps less so. Yet we can only actually work this out by accepting the basic point that jobs, labour, human effort, is a cost of our plans not a benefit.

The rest of this book is an attempt to explore similar economic myths currently widely held to be true and to point to the environmental facts which undermine, prove wrong in many cases, those myths. Our first economic myth is that jobs are a benefit of any plan. The fact is that jobs are a cost and like all costs something to be minimized in our quest, as those greedy yet lazy shaved apes that we are, to get the most we can for the least effort we can get away with.

Just to overdo trying to get this point across, here is an extract from the Green Party Manifesto for the 2010 General Election[12]:

Our energy policy is not just the best for climate change – it also produces the most jobs:

energy source jobs per year per terawatt hour

Wind 918–2400
Coal 370
Gas and oil 250–265
Nuclear 75

They are clearly stating that having more people, rather than fewer, generating the energy we use is a good idea. As one wag (Dave Evans, an internet pseudonym but you know who you are) has pointed out, a recent BBC TV programme showed that if you hook up a generator to a bicycle then each rider will produce 800Wh per eight-hour shift. If people are whipped into doing this 365 days a year then we'll be producing 3.4 million jobs for each terawatt hour of energy.

Yes, it is ludicrous insanity which is rather the point being made.

Notes and References

1. http://www.telegraph.co.uk/earth/earthcomment/geoffrey-lean/7325711/Greening-economies-may-be-the-best-way-to-save-them.html
2. http://bastiat.org/en/twisatwins.html#broken_window
3. An economics professor at California State University at Northridge but he's more interesting than that.
4. http://www.csun.edu/~dgw61315/thetwothings.html
5. Paris Hilton does both talk on the phone and other things in that movie but this isn't quite what is meant here.
6. http://www.wecansolveit.org/content/solution/clean_energy_economy/
7. http://www.ft.com/cms/s/0/372bfcda-ee9e-11dc-97ec-0000779fd2ac.html
8. http://www.jonathonporritt.com/pages/2008/03/the_battle_for_london_mayor.html
9. Either that or their design team still thinks ee cummings is pretty cool.
10. http://www.guardian.co.uk/commentisfree/2008/apr/01/economy.guardiancolumnists
11. An economist and Lib Dem (one out of two ain't bad) who blogs at freethinkingeconomist.com
12. http://www.greenparty.org.uk/assets/files/resources/Manifesto_web_file.pdf

2 Recycling

Bob Spink: *To ask the Secretary of State for Environment, Food and Rural Affairs what estimate he has made of the average time per year spent by a household in sorting and recycling rubbish. [224311]*

Jane Kennedy [holding answer 6 October 2008]: *No such estimate has been made.*

Hansard, 9 October 2008[1]

We all know that recycling saves precious resources: that's why we should all do it. However, as the above quote from *Hansard*, the record of proceedings in Parliament, shows we don't actually know that recycling saves resources: for no one has actually measured the resources that we use in recycling. We have therefore been sold something of a pup.[2]

I would, of course, be among the first to claim that some recycling is a very good idea indeed, that it saves resources. As someone who once made a house's worth of cash by shipping 40 tonnes of old Soviet nuclear power plant off to be made into car wheels I can hardly claim otherwise. It's reasonably easy to recognize this sort of recycling as well: if you can make a profit by doing it you're probably saving resources, for the prices in a market system reflect the value of resources which go into making the various items in that market system.

There is also, clearly, recycling which makes no sense at all in that we all know that recycling of certain items would use hugely more resources than dumping the old stuff and making something new from virgin resources. Cement, say – it is possible to take old and used cement and return it to something which we can use as cement again: most chemical reactions can be reversed if we expend enough energy and effort in doing so. But reversing this cement process takes much, much, more energy than simply digging up a bit more of Portland and baking the stuff. So while we might decide to reuse the building rubble from that well deserved destruction of Brutalist council estates, say, to make roads with, we don't try to recycle it.

Then there's a middle group of things that the market unadorned says are unprofitable to recycle but which, if we look at the wider issues, might well be. This is a result of what economists call "externalities", things which are not encapsulated in market prices. Since they're not already included in these prices then they're not included in the decision-making process of those being guided purely by market prices: they are external to market transactions. A clear case of such externalities would be pollution: if I don't have to pay the price of pumping pollution into the atmosphere then I won't incorporate that cost to others in choking lungs and a despoiled countryside into my prices and everyone choked and despoiled is suffering without my paying for it.

We might find that if I were paying the costs of my pollution, if these externalities were included in market prices, that I would stop polluting. Or that I would pollute less. Possibly, even, that the costs of my pollution are so low as compared to the value of what I'm producing that I'd carry on as before.

It's a common trope that economics simply doesn't take account of these externalities and thus that much of modern economics is wrong. Which is an odd thing to say really as it's modern economics which has firstly uncovered the problem and secondly provided the solution. The uncovering was first done by Alfred Marshall when he was Professor of Economics at

Cambridge around the turn of the last century. So we've known about the existence of externalities and the problems they pose for some 110 years. The solution was proposed by Marshall's successor in the chair, Arthur Pigou, in the 1920s and it is after him that we name this solution, Pigou Taxes (rather oddly the original discussion centres on the effect of too many rabbits in one farmer's fields upon the crops of the next farmer over but it's amazing what examples you can use to get to economic verities).

Quite simply, if my pollution causes costs to other people of £10 then my doing that activity should be taxed £10. The costs I am imposing upon others is now contained within market prices and will influence the decisions that I make. We'll come back to the implications of this in the chapter on efforts to use carbon taxes and cap-and-trade systems to fight climate change. The point here is that to say that economics ignores these points is nonsense. Marshall was the founder of neo-classical economics and the author of the leading textbook of the day. Pigou was similarly, as one might expect of the Professor at Cambridge, one of his age's leading economists and was one of the first academic sponsors of John Maynard Keynes. We're simply not talking about some minority interest out of the mainstream here.

It is also true that we can have positive externalities. These we don't tax, these we subsidize. Just as if people aren't paying the true costs of their pollution we'll get too much pollution, so also if people cannot gather unto themselves the benefits of their activities then we'll have too little of those activities. The usual example given here is something called public goods.

No, public goods are not things which are good for the public, nor are they things which the public thinks will be good for them nor even are they goods which are provided to the public. The meaning is quite specific, public goods are those things which are non-rivalrous and non-excludable. Nasty jargon words, yes, but with solid meanings. Non-rivalrous means that my use of whatever it is doesn't stop you from using it as well. Non-excludable means that it's not possible to stop all these people using it. Think of Newton's equations about gravity here: my using them to ponder why apples fall from the tree doesn't stop you using them to fire a rocket at the Moon. Nor is there, now they've been published, any way for either of us to stop the other doing so.

Such basic science is an example of a public good and is the reason why we subsidize those people in the ivory towers. Because we can't stop anyone from using these equations there's no way to make money out of having discovered them. But if there's no way to make money out of the discovery, why make the effort to discover something? Sure, some people will beaver away at the mysteries of the universe anyway, but, given that we already assume that people are both lazy and greedy, perhaps not as many of those mysteries will be unlocked as we would like. So we subsidize people to do this thinking for us. That we get the World Wide Web (from CERN and Tim Berners Lee), various drugs, alloys and inventions is all to the good and that we also get advanced degrees in "Womyns' Studies" is less so but we must take the rough with the smooth in this life.

Another example of the fine dividing line between goods provided to the public and public goods might be the difference between the National Health Service and public health. My having a hip replacement is rivalrous and excludable so it isn't, in this economist's sense, a public good. But that the country is not ravaged by measles, polio and smallpox is. Yes, of course, those who have had the vaccinations have not benefited from a public good, but there's also this thing called herd immunity. Those who have not had the vaccinations also benefit from the fact that those who have had aren't both falling over dead and stinking up the streets nor infecting them with these contagious diseases. Whether or not the NHS should be tax funded and free at the point of use is therefore an entirely different question to whether vaccination programs should be subsidized from tax revenues. The latter provides a public good and thus should be, whatever one thinks of the merits of free at the point of treatment hip replacements: or of hobbling Worstalls.

One useful way of dividing the knows from the know-nots over public goods is to look at reactions to the existence of copyrights and patents. There are those who deride them as some free market nonsense and that really we'd all be much better off if such information and creation were free to all to use. It's even possible that we would be better off if all were free for use – the copying being done on the intertubes with music and films is probably going to let us know real soon now. But to call such copyrights and patents a free market solution is to miss the whole point of them in their entirety. What they are is an intervention into markets to fix one of the perceived problems of them.

If this book or your song or his invention can simply be picked up and used by anyone, without reducing the possibility of others also doing so, further, if we've not got any way of stopping people from doing this, then we've, well, look up a bit, we've got a public good again. And we tend to think that if people can't make any money out of producing such goods then we'll not get as many produced as we might want. So we artificially create these property rights so that people can make money out of their creations and thus we get more of these lovely creations for us all to enjoy.

This isn't to say that copyright or patents are necessarily the best way of going about this: there are other possible solutions like direct subsidies to those doing the creating. Unfortunately this leads to the sort of novels which come from creative writing courses and writer-in-residence programmes which isn't all that much of a step forward either.

The point of this long digression is that externalities, whether positive or negative, are well ensconced in the basic economist's view of the world and inform our actions and structures in large areas of life. So the concept of recycling being valid for certain things, in certain manners, as a result of the existence of such externalities which need to be compensated for isn't something that the mainstream ignores.

We do, though, come to something of a problem when we consider externalities. It's all too easy for someone to claim that their activity has such wondrous positive externalities that it has a claim on the money vacuumed from our wallets by the tax system: this is how we end up with advanced degrees in Womyn's Studies and writer-in-residence programmes. It's also fairly easy for people to say that this or that negative externality is so expensive that really, people shouldn't be allowed to do that. This is how we ended up with no wind farm in Nantucket Sound where Teddy Kennedy liked to go sailing: his being able to sail was of more value than electricity for tens of thousands as Teddy Kennedy pointed out to us all. We thus need to have some method of arguing through these claims, of weighing benefits against costs.

Which we do have of course: the weighing of costs against benefits is called a cost benefit analysis and as we've seen in the first chapter we know how to do these. They aren't perfect, they come up against the bounds of human knowledge and that can never be perfect. It's what Ludwig von Mises called the socialist calculation problem (or for those just against planning in general rather than socialist planning in specific, the economic calculation problem). There are simply so many people with so many different desires that we cannot really ever know what would best please everyone. OK, we can make some fairly good stabs at it at times – not dying a horrible death in the aftermath of a nuclear war we can assume to be a reasonably widely held wish. The precise width of shirt collars is one we probably can't solve centrally as the knowledge of different desires simply is not (and in Hayek's version, cannot be) discovered centrally.

But despite these constraints we have to proceed as best we can.

As we saw in the first chapter we do need to make sure that we put all of our costs on the costs side just as we should make sure that we put all of the benefits on the benefits side. This is what brings us to that quote from *Hansard* at the top: when we talk about domestic recycling of waste, it's not just that we don't know one of the major costs, it's that no one has even bothered to find out what that cost might be. The cost of the labour which we have to do inside the household in order to sort rubbish so that it can be recycled. We therefore don't know whether recycling does indeed save resources or not.

On the benefits side of recycling we have the obvious point that some of those things we do recycle save us money. Aluminium drink cans are a good example: the energy required to make new aluminium from bauxite (the process is bauxite to alumina and then alumina to aluminium metal, that second step using so much energy that we ship the alumina around the world to where energy is very cheap, like hydroelectric sites in Iceland or Quebec, even build great big new dams just to drive such aluminium plants) can cost as much as $900 per tonne. If we can save that energy just by collecting old cans and avoiding that process, just melting them down to make something else, then great, let's do that. Similarly copper piping, steel, old cars and so on. This is our first form of recycling mentioned above: as we can make a profit from it we're clearly saving resources.

We might also say that if we throw food into landfill then it will rot. When it rots it gives off methane (at least, if it rots without there being oxygen around) and methane is a greenhouse gas. Thus putting food into landfill brings closer that day that a Pacific island sinks below the waves. This is a cost that we are imposing on others, one which we are not paying for. It is also true that since the 2004 Landfill Act we have to collect this methane from rotting food and indeed we do so:

burning that gas provides some 30% or so of the UK's renewable energy at present. We can go on as well: when we burn the methane we produce CO_2 which is itself a greenhouse gas. However, methane does some 23 times more damage than CO_2 so we at least reduce the damage done, while at the same time getting some energy (energy that we would probably have produced some CO_2 to get, in another manner, anyway).

However, what if we didn't stick the food at the bottom of a pit? What if we composted it instead, used a wormery perhaps? We would get some useful stuff to put on the garden at least...but we'd also quite possibly put even more greenhouse gases into the atmosphere. For worms fart you see.[3] Sadly, such worm farts are nitrous oxide which is itself a greenhouse gas. A greenhouse gas which is 296 times worse at warming up the world than carbon dioxide is and thus over ten times worse than methane. Now, that nitrous oxide comes in much smaller quantities is true (there is only so much flatulence a worm is capable of even if it is digesting the lentil stew from a rigidly environmentalist and thus vegan household), but researchers at the Open University think that the total greenhouse effect of the worms would be about the same as the effect of landfilling the same food. More, then, than landfilling and burning the captured methane.

Some glass is worth recycling in the UK. While the basic material to make glass is common (it being sand, yes, simple sand from a beach will do) it does require a good dose of energy to melt it and make the glass. Melting already used glass uses less energy. But even WRAP (the Waste & Resource Action Programme, the government advisor on the matter) says that green glass is not recyclable in this manner in the UK. This is because most green glass is used to bottle wine: something we don't produce a lot of in Britain but something which we are noted for importing quite a lot of. There's no real use domestically for the green glass that is so carefully collected and it's crushed up and used as hardcore to underly tarmac on roads.

WRAP also admits that this almost certainly emits more greenhouse gasses than simply throwing the bottle in a hole in the first place.

Perhaps the decision has already been taken for us though? Maybe we don't have enough holes in the ground to dump all this rubbish and therefore we must recycle it? Given that the UK digs holes with a capacity of around 110 million cubic metres (mcm) each year and we've got a volume of rubbish of about 100 mcm each year this isn't true either. We need to get our sand, gravel, clay and so on from somewhere and we're also fairly keen on filling those holes up when we've finished and rubbish isn't the worst thing we could put down there. We do have a shortage of holes that we're allowed to put rubbish in, this is true, but that's simply because government won't licence many of the holes we do dig as landfills.

The reason for this refusal to licence refuse dumps is obvious even if it's not all that sensible. Quite simply, we've been told by the European Union that we must recycle more of our rubbish. One claim is that it is good for the environment that we do so, that it saves resources, and that's certainly the public statement even if that's still something that has to be proven rather than asserted. The actual mechanism seems to be that it would be a breach of Single Market rules if we were to carry on gaily throwing detritus into holes in the ground. For, you see, there are countries which don't have our good fortune in having lots of holes to throw detritus into. Holland for example, where mining for sand or gravel is likely to lead to the discovery of the North Sea again. For us to be using holes would be unfair competition against those who have no holes to use and thus must be *verboeten*.

Yes, it does sound ridiculous but then this is the European Union. The Jams, Jellies, Marmalades and Sweet Chestnut Purees regulations (as revised, 2004) insist that carrots are to be defined as fruit. There is little that should therefore surprise any of us about the justifications put forward by that august body so essential to Germany not invading France. Again.

Our cost benefit analysis is getting quite complex, as we can see. This is why we leave such things to government itself to do rather than trying to tot it all up on the back of an envelope before working out whether to put the teabags in with the baked beans tin. At least, this is why we should leave it to government if we're willing to make the assumption that said government is going to do the analysis properly, that they'll consider all of the relevant issues. Which, sadly, as that quote at the top of the chapter shows, they haven't.

The current analyses of whether or not we should recycle domestic waste leave out an important cost: the time that we have to spend in sorting through the waste in order to prepare it for recycling. There is some recordable amount of time that it takes to do this: it isn't just a matter of making sure that you put the glass in with glass, tins with tins and food with food. Jars must be rinsed out, plastics separated (some places insist that the plastic of a window envelope be separated from the paper, others that TetraPaks must be disassembled) and so on. Yet as that Parliamentary reply tells us, the government has no estimates of how much this time actually is nor how much it costs us to supply it as the law insists we must.

That particular question was asked at my urging when I was working in the press office for the UK Independence Party (UKIP and yes, my being both UKIP and Adam Smith Institute is going to pose difficulties in getting some people to take these arguments seriously, isn't it?). I also got Godfrey Bloom MEP to ask the same question in the European Parliament and he got the same reply: we dunno. In fact there are very few studies indeed which attempt to track down how much time preparing to recycle takes the average household. The only one I've been able to find which has any rigour to it at all (and which isn't clearly entirely biased one way or the other) is from the Seattle

Public Utilities who did a survey along with the University of California at Davis[4].

For a simple sorting programme, metals, glass, paper, everything else, they estimate that it takes 16 minutes per household per week. For a complex programme, including food and garden wastes, the sort of composting and so on that we're all being urged to do now, more like 45 minutes a week. I agree, this doesn't sound like all that much, trivial in fact against saving the planet. But there are 24 million households in the UK and 52 weeks in the year. That's some 312 million to 936 million hours of labour a year that's going into just sorting our rubbish so that it can be recycled: a fair old whack of effort going in we might say. If you prefer, at the average full time working year of some 2,000 hours that's 150,000 to nearly half a million full time jobs being spent just on not landfilling stuff.

(A quick note about numbers throughout this book. Very few of them are exactly accurate: whether there are 24 million households or 24.3 million makes no difference to the arguments. Similarly, 37.5 hours a week for 52 weeks a year is 1,950 hours but using 2,000 changes little. It's simply boring to be striving for accuracy as long as the first few digits and the order of magnitude are correct. Everything therefore gets rounded: the top estimate above of full time jobs is 480,000 if we're to be accurate but half a million is close enough for our purposes: or, if you prefer, twice the size of the entire Armed Forces.)

We could at this point just go back to chapter 1 and point to opportunity costs: what else could we be doing with half a million people for a year that's worth more than making sure we still have empty holes? Or we can move forward and apply some costings to the various parts of this system and see what the balance is.

Way back in 2001 the Prime Minister's Strategy Unit started to worry about the costs of the entire waste disposal system. Their report was called "Waste Not Want Not"[5] (and we should note they were talking about England only but see above for our required accuracy of figures) and they pegged the then cost of the system at some £1.6 billion. They also predicted that by about now the costs would be near £3.2 billion and so there should be a plan to reduce the rise in costs. Recycling was hit upon as the way to do this. All of which is entirely excellent of course as long as recycling does in fact cut costs. But we've got that 300 million to 900 million hours of labour which we've got to account for in some manner before we can decide whether we have in fact just cut costs.

So now we've got to find a way of estimating what that labour is worth, what value should we assign to it. The easiest thing to do is to say that heck, it's 5 minutes here and 10 minutes there and so it doesn't matter. Forget it: some people do indeed say that. Or we could say that if people weren't doing the recycling they'd be doing something, sure, but they wouldn't be working, so we can't assign a monetary value to that time. Some people do indeed attempt to say that as well. We might even say that everyone is just overjoyed at the idea of saving Gaia, so are happy, ecstatically so, at being given this opportunity to do so. Yes, I've seen people say that too but if this were so then there wouldn't be any need to fine those who don't recycle: for there wouldn't be any who didn't. That we do indeed have fines, that we have laws that insist that you must recycle, means at the very least that those making the laws don't believe that people are so ecstatically happy to do it all without compulsion.

Just to try and hammer home this point that time has a value over and above whatever we produce with it: very few of us go into that long dark night complaining that we've had too much time here. In fact, we know how many people do so, at least, those who think so so strongly that they do something about it. The number who commit suicide each year varies between six and seven thousand. While this may be a harsh calculation that number is therefore the number of people who think that life is too long: the other 65 million of us seem to find enough things to occupy our time between birth and that hopefully delayed death. Someone taking that time off us to do their bidding thus should be assigned a cost. We do not run the entirety of society for this small number of people (although we might perhaps pay more attention to mental health inside the NHS). Well, actually, we do, that 6 or 7 thousand is roughly the size of the professional political class in this country and we do seem to run the whole thing as they would wish it were: I'm sure you have days, as I do, that getting the suicidal to swap with the politicians would be a good idea. The former won't do any worse than the current incumbents and the latter will be thankfully gone from our lives.

This idea of free time which can simply be diverted to whatever the government of the day desires, without people having to stop doing other things to make room to do what they're told, also violates the basic economic concepts of how people do use their time. We've all heard of this work/life balance thing, and being told you've got to divide the eggshells from the kitty litter is at least an imposition of work on that life part. Economists go further though and divide our time up into four different parts, measuring each through something called "time use surveys". There is personal time, leisure, household production and market production hours. The distinctions can blur a bit at the edges: personal time is defined as those things which you cannot get someone else to do for you. No one else can sleep for you, take your shower or eat to keep you nourished. The blurry edges come where, say, sex is concerned for you most certainly can turn that into a market or cash transaction as street corners across the nation show us. Fortunately we British don't spend enough time on such grubby matters for this particular blurring to be a major problem. Household production is those things which we do, not unnaturally, in the household. Cooking, cleaning, fixing the car, clearing the gutters, mowing the lawn. Market production is us going out to work and earning money or, more rarely, us doing something at home which we then sell: the cash part of life if you like.

The lines between household and market can blur as well: if we cook at home that's household production but if we get in a takeaway then that's someone else's market production that we are consuming. My cleaning my bathroom is household, my wandering out with a mop and bucket to clean bogs for a living is market. The balancing number, the one that gets us from personal, household and market time to the 24 hours of a day, is leisure. And yes, it really is true, whatever people bleat about that work/life balance, that leisure hours have been increasing for the past couple of centuries. We do not have "ever longer working hours" we have, in total, ever shorter ones.

What has been happening with working hours is that female market hours have been rising. This is essentially the result of the economic emancipation of women: we can argue merrily about how this happened (less patriarchy, the pill, the feminist movement, the way that modern service industry doesn't value male musculature like the old manufacturing or agriculture did) but that is what has happened. Male market working hours have fallen at the same time: the usual suspect is that as we've all got richer we've taken some of that increased wealth as leisure rather than higher incomes. For both men and women unpaid hours of household production have fallen dramatically: this is essentially a result of increased technology. You only have to boil nappies now if you really want to, a gas stove is so much more convenient than a wood fire, washing machines and tumble driers have replaced mangles. For men, cars are more reliable and need less maintenance, veg patches have been replaced by the supermarket and DIY is (for some unfathomable reason to me) something people do by choice, not the only way you're going to get that door fixed. The fall in female household hours more than covers the rise in female market hours.

Total working hours have been falling for at least the past century for both men and women, leisure hours rising. Now, with this recycling requirement we're demanding, by law, that people should do more household, unpaid, production hours at the cost of cutting into their leisure. Yes, it might indeed be true that we're only insisting they give up a couple of minutes a day of the average several hours of TV watching. But it's one of those things that economists are really rather persnickety about. People's time has a value: the value those people themselves put on what they could be doing instead of whatever it is that you are forcing them to do.

No, this value is not dependent upon their using this time to produce something which others value, like working more or producing more. People quite rightly value the time supping a fresh pint, playing with the children, some strange types have even been known to enjoy staring at the sunset hand in hand with their *inamorata*. If we insist that they must do something else other than these things which they themselves enjoy and value then we've got to come up with some method of providing a valuation of that time that we insist upon. Only if we do that can we work out whether what we're insisting they do is actually worth it: think a moment, if you could demand 900 million hours of free labour for your special little project you could make just about anything sound profitable, couldn't you? But it is only profitable, it only adds value, if whatever is produced by your project outweighs that labour you're getting for free.

Fortunately this has been dealt with by a couple of Nobel Laureates[6] just recently. Joseph Stiglitz and Amartya Sen were asked to advise Nicky Sarkozy over in France on alternatives to GDP. A worthy idea, for GDP isn't all that good a measurement of human progress, with just one proviso.

There are any number of reasons why GDP isn't perfect, most of them acknowledged by Simon Kuznets when he invented the concept. We're not measuring the distribution of what is created, only its creation. We're not measuring household production, only market production. We count cleaning up pollution as adding to GDP but not the losses from that pollution in the first place.

That last is a common complaint these days: cleaning up that BP spill in the Gulf of Mexico will indeed be counted as an increase in GDP while the losses caused by the pollution in the first place will not be. So it would appear that we can all get rich by spraying oil around the place. However, including the clean-up process in GDP does make sense: the reason we're doing it after all is because we think that cleaning up such a mess adds value. The fault is in GDP not including the original damage. Which is where the really good thing about GDP comes into play.

There are any number of possible ways of measuring economic output. Gross Domestic Product is only one. We can talk about Net product, or gross National product or net national Income, all making adjustments for this or that part of what is wrong with GDP. For example, if we were to measure Net Domestic Product (NDP) then we would be capturing the losses from the original pollution for by measuring net, not gross, we take account of depreciation and the using up of scarce natural resources: the environment being one of those of course (one factoid is that when you do measure such consumption of natural resources then among the rich countries it is Norway which is depleting its environment the most. All that oil and gas being pumped up to pay for the icy social democracy you see). But that good thing about GDP is that it is much easier to calculate than that alphabet soup of other measurements. We can crank it out (subject to later revisions of course) a few weeks after the end of whatever period we want to measure. We can't really calculate the Net part until all of the companies have reported their annual results, 12 to 18 months later, and we see what depreciation was. This would be even more like using the rear view mirror to guide our driving than the current system: not a great advance.

It's at about times like this that Sir John Cowperthwaite seems so clever. While Financial Secretary (ie, British civil servant sent out to keep the natives in line) to Hong Kong he refused to let anyone collect GDP statistics. People would only use them to do something foolish.

But having agreed that despite GDP's faults, it's still pretty good because we can at least calculate it, our duo of Laureates investigated how we could make the measure better. There are two points relevant to our argument here. The first is that they are insistent that when measuring work we should not simply measure market working hours. Rather, we should be measuring the residual, leisure hours, for that is what is important to the human beings doing the working and the leisuring. Whether we work 10 hours at the factory scrubbing grates for sale and one cleaning the house, or one working in the factory and 10 scrubbing the stove doesn't matter all that much. What we're all interested in is how much time we get to sit down and read a book (umm, OK, in the modern world perhaps watch a soap opera).

This supports the point made above, that work we are forced by fiat, to do in the house, is still work and must be accounted for as a reduction in our leisure.

Their second relevant point for us is that they try to put a value to the output that comes from that household or domestic production. The political point behind this is, well, rather political really. It's a general truism that lower tax societies like the US tend to do more market working hours than high tax countries like France. However, those in France don't have notably more leisure hours than Americans for they do more household production. Nothing very surprising here: in a low tax environment you get to keep more of what you make and thus you'll happily work for the man, collect your money and buy what you want. In a high tax environment there's a bigger bite taken out of what you earn and you're more likely to decide to do more things for yourself, inside the home, where the taxman doesn't get any of it.

But there's a side effect to this: GDP doesn't count such household production. So while people are indeed eating mother's hand-sliced and toasted Croque Monsieur rather than a paid-for Grilled Cheese, Americans seem richer because they have bought that toastie in the market, not made it at home. This of course would never do, the hyper-capitalism of *Les Etats-Unis* leading to a higher recorded standard of living of *La France Profonde*. So, the report suggests a manner of measuring the value of the output that comes from all of this home labour. We should measure it at the "general undifferentiated labour rate". Or, as it will be in those countries which have one, the minimum wage.

This idea of measuring output by assigning a value to the input is indeed a little strange but it's also not unusual. We often come across bits and pieces of the economy where we don't really know what the value of the output is. Government being a good example: yes, it's lovely having courts and police to deal with criminals, valuable to have hard men who would do violence on our behalf to Johnny Foreigner in the night. But it's very difficult indeed to put an actual value on this output of a criminal justice system or national defence. So, in fact, we say that the output is the same as the input: the value is the same as what we spend on it. And this is how we measure government's contribution to GDP: it must be worth what we spend on it. That this tells us that if we raise MPs' salaries MPs are therefore more valuable to us is just another one of those annoying failures of GDP itself.

So there's nothing unusual about this method of valuation: yes, we can see the problems with it. My wife's Shepherd's Pie is indeed an addition to the value of the output of the nation, my own essay at the same recipe undoubtedly a diminution of said value but we just accept that this will all even out and we value, now, household production at the value of the labour that goes into it. Or, as I write, £5.91 an hour in the UK.

So now we have three important things: we've an insistence that hours spent not in leisure must be valued as an input into whatever scheme it is that we're insisting must happen. We have an estimate of the hours that we are insisting must be put into the sorting of domestic waste for recycling. We also finally have a cash value that we can apply to each hour that is so spent sorting. Our 300 million hours at £5.91 gives us a cost of £1.8 billion a year, 900 million hours of £5.3 billion.

Now remember, right back at the beginning, our original problem was that waste management might cost as much as £3.2 billion at some point in the future. To avoid this cost we seem to have thrown as much as £5.3 billion of labour at it: it isn't immediately apparent that we've saved resources by doing this, you know.

Even if we say that there's a £2 billion revenue from selling recycled items (the 2008 figure) it's still not entirely apparent that we are saving resources in this manner. For the waste disposal system itself still costs something, plus that cost of labour, and we get the revenue plus the reduction, whatever it is, in the cost of the waste disposal system.

What we've really found out is that we don't know whether we're saving resources or not through this recycling, because no one is taking account of the labour that must be used to sort before recycling.

Having come out fighting and (at near wearisome length) shown that recycling doesn't in fact save resources, let me backtrack a little. Let's assume that we are all indeed good little

Gaia worshippers and that it really is a sin against the planet to be putting last week's chicken tikka into a hole in the ground. That we really should recycle envelopes, that green glass should be made into roads. What we would really like to know is what is the best way of doing this? What method can we use which performs this religious task at the least expenditure of resources?

We could imagine two different methods here: the first the one described above, where every house has a slop bucket, nine different containers for rubbish (as at least one council is currently insisting) and wormeries proliferate across our green and pleasant land. Or we could dip into a little bit of Adam Smith and see if there's another method possible. One of Smith's great points being the value of the division and specialization of labour.

So much so that the quote on the back of a £20 note is currently "The division of labour in manufacturing: (and the great increase in the quantity of work that results)". I've had a snarl at some poor lady in the printing department about that quote (yes, I am like that) for the meaning of "work" has rather changed. We would now associate it with the effort put in rather than what Smith himself would have said, the quantity of finished work that results from a given effort put in. But we could imagine a system in which households just dump everything in the one bin in the ancient and historical manner. Then we use some centralized place, perhaps even coin a name for it, like recycling centre, factory even, to separate all of the various different items. We would assume, at least as a first step, that specialists would be more efficient at the separating and sorting. We could also mechanize some of the steps: various options such as magnetic separation (iron, steel, anything nickel etc), air flotation, even a modicum of hand picking where mechanization isn't possible.

The process itself might be more efficient: we'd also be running fewer collection rounds meaning an environmental

saving there as well. Indeed, this could be like the place my mate
Keith used to work in in Nottingham. Pour an undifferentiated
domestic waste stream in at one end, collect sorted waste at the
other. Now we have made an assumption here: a possibly
unwarranted one. That our specialists are going to be more
efficient at this than households, that we'll use less labour, fewer
resources, to get to the same end, and desired, result, properly
sorted domestic waste. So we'll need to work through whether it
is actually more efficient . . . yes, you can see this coming again,
can't you: a cost benefit analysis.

Clearly if we're going to create a class of waste-sorting
operatives we'll need to be paying them wages. If the unions get
anywhere near it we will also need to factor in huge pensions
and sick days for a sniffle: if the bureaucracy gets a sniff no doubt
we will need outreach, diversity and gender allocation advisors
as well. But it still might be more efficient: except, from all of the
above we know that we're not counting the value of the labour
that goes into the current system. Which is why we must count
that value for it's possible that a different way of organizing the
same process could require less labour. A better work/life balance
that is, less labour required and thus more leisure available to be
spread around the population of the country. Something which
is desirable you know, reducing the amount of work we have to
do, increasing the amount of time we have to pursue our own
desires.

That's why we have to assign a value to the labour used in
recycling. Not so that I can make (however enjoyably) snide
remarks about whether or not we are saving resources by
recycling. But because we don't know, cannot know without
measuring such labour, whether we are using the best method
to recycle.

Finally, on this subject at least, for those who think that
time should not be counted as a cost at all. This is from George
Monbiot looking at the case for high speed trains. George isn't

taken with this method of calculation but the government clearly is[7]:

> The cost-benefit analysis (which the government calls "the business case") produces benefits of £32.3bn. The department concludes that the scheme has a benefit-cost ratio of 2.7. But where did the £32.3bn come from?

> Almost all of it is money deemed to have been saved by reducing travel times.

Ah, you see? When the government wants to persuade us of the case for something they're quite happy to include time saved as a benefit, a benefit with a cash value assigned to it. Which means, ineluctably, that time spent is a cost, a cost with a cash value to be assigned to it. To say otherwise would be to imply that we are ruled by people who are just making it up as they go along.

Unthinkable, surely.

Notes and References
1. http://www.publications.parliament.uk/pa/cm200708/cmhansrd/cm081009/text/81009w0004.htm#0810096000521
2. http://www.theecologist.info/key27.html
3. http://www.telegraph.co.uk/earth/earthnews/3299669/Wormeries-may-add-to-greenhouse-gases.html
4. http://www.weeklystandard.com/Content/Public/Articles/000/000/006/603wxcce.asp?page=2&pg=2
5. http://www.cabinetoffice.gov.uk/media/cabinetoffice/strategy/assets/wastenot.pdf
6. Yes, I know, it's the Swedish Bank prize in honour of Alfred Nobel, not a real one.
7. http://www.guardian.co.uk/commentisfree/2010/may/17/high-speed-rail-policy-carbon-emissions

3   Growth

Anyone who believes exponential growth can go on forever in a finite world is either a madman or an economist.

Kenneth Boulding

You'll have seen this quote before no doubt: had it thrust at you perhaps as the perfect refutation of this idea that we can have continual economic growth. For, look, see, a famous economist has said that growth forever isn't possible! And Ken Boulding was indeed an economist, a very good one, if a little to the heterodox side of things.

Heterodox here, when applied to economics and economists, can mean anything from those who look at the world with a slight slant, at an angle, to the entirely barking mad. Boulding was very definitely at the interesting, perceptive but at an angle end of that spectrum as opposed, say, to the more rigid Marxists of our time. The barkingness of Marxism shouldn't really have to be explained now that we've all seen the rubble of the various attempts to introduce it into the real world, but just two little examples:

The first is that Marx warned of what would happen if the capitalists ganged up on the workers: he knew very well that what led from rising productivity to rising workers' wages was that capitalists would compete among themselves for access to that labour, the rising productivity of that labour and thus the higher profits that could be made from it. That competition for access to the profits that could be made is what would bid up wages. And we see, in market economies around the world, as labour productivity rises, the workers' wages rising.

What his followers then did as they built the Soviet Union was to make the State the sole employer of labour. Stalin explicitly laid out the implications of this: with no competition between employers then the capitalist, the state, could hold down wages so as to increase the profits made from labour. This was, at least they said it was, done in order to create the capital necessary to build communism: something that sure took a long time acomin'. The net effect, of course, was that an avowedly Marxist system ended up doing exactly what Marx had been warning against. Building a monopoly of capitalists in order to reduce the workers' wages and thus boost profits.

The second is this insistence that any profit at all is stealing the sweat from the worker's brow, removing some of the value he has created, unfairly, immorally removing some of that value. Stealing it, in fact, which is why every capitalist should be hanging from the lamp posts in a class version of "Strange Fruit". A very strange idea indeed, for isn't the worker also appropriating to himself some of that surplus value created by capital?

Think of it this way for a moment: the farmer digging the vegetable patch with his hands is not being all that productive. When the capitalist provides him with a capital good, to wit a spade, one, he becomes more productive. The capitalist gets some of that extra production in return for the provision of the spade. Our standard Marxist analysis says that this is theft by the capitalist and that all of the extra production should belong to the labourer. But hang on a minute: the capitalist doesn't get all of the extra production from the provision of the spade. The farmer gets some of it too: so the farmer is equally exploiting the capitalist. The labourer is getting some of the surplus production from the employment of capital.

An economic or philosophic system based on ignoring either historical reality or this philosophic conundrum is of course barking mad and thus so are the purer and more rigid of Marxists. Ken Boulding was not so but in this specific case he was wrong although wrong in an interesting manner. For he was wrong in a way that economists normally are not: he made the assumption that the physical world is the defining limit of economic growth, something which simply is not true.

That resource availability is a limit upon economic growth, this is certainly true. Economics is, after all, the study of the allocation of scarce resources. The scarcity of those resources (whatever they are, labour, fresh water, copper) limits what we can do. We've already looked at opportunity cost which tells us that we cannot do two mutually exclusive things with the same resource: our example was that we cannot both ski in the mountains and surf at the beach at the same time. The same is true of our piece of copper: we cannot use the same kilogram of it to wire up a telephone in China and also to pipe water into a house in Soweto. So certainly, yes, the scarcity of resources does indeed impose limits on the size of the economy, upon economic growth, at any one time.

However, what is true at any one moment in time is not true over time, or not necessarily so. For two reasons that economists continually try to point out. The first is substitutability: anything and everything is substitutable. This does, at extremes, rather become farcical: death is the outcome, as we know, of the complete absence of food. Death is therefore a substitute for food, however unwelcome a one. But staying on the reservation of reasonableness there is a point: if we have no copper we can run fibre optic to the house in China and plastic pipe to the one in Soweto. Which of the four actions we will choose will depend upon the relative prices (which will reflect both the scarcity of the resources and the value we get from their employment) of the resources we must use and the possibility of substitution between them. In this case, copper is not the limiting resource of either potable water in South Africa nor of being able to call the mother-in-law in China.

The second is that we don't so much consume resources as create them. Yes, this does indeed cause a lot of headscratching but it is still true. Yes, absolutely, there are only so many copper atoms on the planet. There are only so many iron ones as well, so clearly we cannot use more such atoms than there are actually available to us. But that number which is available to us is fixed, not by (except at a truly enormous extreme) by the number on the planet, it's fixed by the technology we have to make use of them. Imagine, back in the Dreamtime, wandering past Mount Whaleback in Australia. It's an interestingly red mountain, that's pretty much it. Once the blonde despoilers of Gaia turn up with their knowledge of turning haematite into interesting ways of killing people it becomes tens of millions of tonnes of iron ore on the hoof. We have created the resource by having the technology to use it: so much so that the Mount has been turned into Hole Whaleback as we continue to scrape the 68% pure iron oxide out of it.

OK, iron is a pretty old technology, several thousands of years old at least, so perhaps this is something that we've already done, reached the limits of? No, not really: back in 1968 a new method of getting copper out of dirt (it's all dirt until we've got the technology to get what we want out of it: it's the technology to do so that turns dirt into ore) called SX-EW (no, don't worry, it's boring what it means) was invented meaning that what had been dirt with copper oxides and sulfides in it now became copper ore. We had created more copper, made a resource, by advancing the technology to get it.

It's not limited to 40-odd years ago either: my project this summer, after I've handed in this manuscript, is to run around applying a variety of technologies to a variety of dirts and see if I can turn them into scandium ore. Don't worry, no one else has heard of scandium either but it looks like it might be very useful in making a certain type of fuel cell, one of those technologies that's going to wean us off fossil fuels. If successful, I and my muckers will have created scandium, made available what was not available before, created a resource. No, of course we haven't made scandium atoms (although that can be done in a nuclear reactor) but in the sense that economists use the words "create" and "resource" we will have done (might do, hope to do, whatever).

Of course, this is all economic nonsense, jargon, no one outside the ivory towers really believes that, do they? Well, actually, everyone who really thinks about this subject does believe it. Arguments from authority are of course logical fallacies: an argument should stand or fall on its own merits, not according to who makes it. Yet in this whole discussion about climate change we do have an authority, the Intergovernmental Panel on Climate Change, the IPCC. We are advised to take them as the scientific consensus, abjured from calling it all nonsense. Which makes this line from one of their reports so interesting[1]:

> Energy and mineral resources are abundant in this scenario family because of rapid technical progress, which both reduces the resources needed to produce a given level of output and increases the economically recoverable reserves.

If the very people who tell us that climate change is happening, the very people we rely upon for the whole insistence that we must do something about climate change, are telling us that technological progress increases the availability of resources then, well, when we're thinking about what we should do about climate change we really ought to take note of what the people telling us about climate change are telling us.

We create resources by inventing the technology that allows us to use those resources we've just created. So the limitation of resource scarcity upon growth is not the resources themselves, it's the technologies we have to make them available to us.

We'll look at the underlying economics of climate change in more detail when we come to globalization. Here it's just worth pointing out that the entire process, everything, Rio, Kyoto, Copenhagen, windmills all over the countryside and slops bins in our kitchens, relies upon a series of economic models listed in the Special Report on Emissions Scenarios (SRES) from which that quote is taken. In those models, in 2100, the global economy is between 5 and 11 times larger than it was in the 1990s when the models were constructed. Yes, we have problems with the atmosphere, the temperature: but we do not have a problem with a shortage of other resources: so if we are to believe the forecasts of problems with the atmosphere and the temperature then we really also ought to be acknowledging the other things that they say. Resource constraints are not going to be a limit on economic growth over the next century or so. Or rather, not going to be the binding limit upon such growth.

However, the biggie, the killer point if you wish, that economists use to point out why the physical world is not the limit to economic growth comes from the very definition of what economic growth is. This is where we can say Boulding went wrong. To avoid being accused of cherry-picking my definition, allow me to use one from the nef, one they present in their report which introduces us to the nine billion tonne hamster as an image of the economy (what are they smoking over there?)[2]:

> Hence, an economy is said to be growing if the financial value of all the exchanges of goods and services within it goes up.

That's not a bad definition, it's close enough to the academic one for us to use it. We tend to measure the size of an economy by Gross Domestic Product (as we've already discussed, this isn't great but it's good enough) and that in turn is the value of goods and services produced in that economy. Growth in that value of goods and services is thus growth in the economy, economic growth.

We might want to take a little detour though, a definitional one, and ponder on how that value is defined. Their use of the word "financial" carries with it, in these days when many would happily hang all financiers, something of an overtone. Yes, we do mean that GDP is measured (as again we have already seen, excluding household production for example) using only the market, the cash, part of the economy. But that is not to say that that value is determined by those financiers: far from it, the value is the often arbitrary, always subjective, value placed upon things by consumers. The value being measured is the value we are willing to pay for something. Not the value of the resources that went into it, not some idealized or "real" value but simply what John and Jane Smith are willing to pull out of their wallets in exchange for whatever it is. We, the consumers, define that value by our decision to purchase or not purchase it: that's what is meant by "financial value" here.

We could look at this one way, the way in which some do. Imagine a physicist looking at the world around him.

Ah, another detour: it's not much of a surprise when people apply their knowledge of a specialist field to another subject that they're considering. We would be deeply unsurprised at an engineer looking at the economy and considering it all to be a matter of hydraulics: to a certain type of engineer everything is hydraulics. And while just about everything does indeed have elements of hydraulics in it, thinking of everything in only such terms tends to miss some of the subtleties. I'm sure you can describe sex purely in terms of hydraulics for example (certainly you can with erections and labial engorgement) but that is, I can't help feeling, to miss some of the point and glory of it all. In fact we did have an engineer who looked at the economy as a matter of hydraulics and he gave us Social Credit as the answer. All great fun but he'd missed the very point that such as Hayek were making at about the same time: you cannot have the detailed knowledge of the economy in the same way that you can of a set of pipes and tubes. Thus an economic system which depends upon such detailed knowledge cannot work as you cannot have the knowledge to make it work.

Similar forays into economics came from a brace of Catholic poets, GK Chesterton and Hilaire Belloc. They ended up giving us Distributionism which is essentially a rehash of what Mussolini said he was going to do: no, we're pretty sure that's not the right answer either. Despite Phillip Blond's "Red Toryism" being based upon it.

But back to our physicist: on looking at the economy or economic growth he might point out that goods have to be made of something, services certainly use energy in their creation and delivery. As there's a limit to what can be made into goods and a limit to energy then there must be a physical world limit to economic growth. But our physicist is making an error: we shouldn't be too hard on him for doing so for he's making the same one Boulding was. The economy is defined as the value of goods and services: there's no physical world limit on the value that can be created, even if there is one on either the materials we can add value to or the energy with which we can do so.

To spiral off to the side again, the problem here is that those viewing economics through the lenses (or perhaps we might say prejudices) of their own knowledge are simply unaware of all the caveats that economists place upon what they say. No, no one is saying that there are no physical limits to growth. Physical growth is indeed limited by the size of the environment in which you're trying to have that growth. But economic growth isn't physical growth, it's growth in value.

As an illustration of this, think of the argument put forward by some of the relatively brighter creationists (and sadly, given that we're talking about believers in creationism, it can only be relative brightness). The Laws of Thermodynamics prove that evolution cannot happen. While CP Snow gave us the understandable version of those laws (you can't beat the house, you can't break even and you can't quit the game) our Bible Toters have got at least part of it right. Systems trend towards entropy and evolution, the production of ever more complex beings over time, certainly doesn't look like any trend to entropy. But as any physicist will tell you, most especially when he sees Genesis being waved in argument, the Laws of Thermodynamics only work in closed systems. Earth is not a closed system, there's a large nuclear furnace some 90-odd million miles away pumping in energy every moment of every day.

No, the existence of the Sun does not prove that God didn't create us all as his special little creatures (although the existence of *The Sun* has some wondering) but its presence does disprove that one specific argument offered in support of the contention. So the engineer looks at economics and says it can all be calculated like water pipes: not knowing the bit that comes with the economist's secret decoder ring that we can never have enough information about the economy to calculate it all. The physicist tells us that as there's a limit to the material and energy from which goods and services can be made there's a limit to the economy. Not quite getting that it's the value we're measuring and there is no physical limit upon that.

To offer one example of how this might work in practice: Solar cells, those things that we're all hoping will save us from climate change. They're made from highly pure silicon metal. We get that metal from beach sand. Getting from beach sand to the large ingots of that highly pure metal takes a great deal of energy. In fact, a large amount of the embedded energy in a solar cell system (something we do slightly worry about as we're not sure that using more energy to create an energy creating system than the energy we're going to get from that system is a good idea) comes from that very part of the process. Once we've got our ingot we slice it and then we start to draw (or etch, print, different ways of doing this) the circuits on it that turn it from a silicon wafer into a solar cell.

Now, imagine, go on, just imagine, someone comes up with a way of slicing that ingot more thinly. We can thus get more wafers, more cells, from the same ingot. We have, with no greater use of resources, just increased the value of our production process. We've just had economic growth without using any more resources. In fact, Applied Materials has been doing this for some years now: the wafers are about 80 microns thick, down from a typical 200 micron or more only a few years ago. We're getting more than twice the value out of our ingot of metal than we were at the turn of the century (actually, these things move so fast that it's more like 4 or 5 times).

We can now have more solar cells made from the same raw materials, how lovely, but for the point of the argument here, we can say that either we are creating the same economic value with less resource use or that we're creating more economic value with the same resource use. Meaning, clearly, that resource availability isn't the binding constraint upon the creation of economic value and therefore is not the binding constraint upon economic growth.

Which then means again that if resources are not the binding constraint upon economic growth then continual economic growth within a finite physical system is entirely possible. Just as long as we continue to find new ways of adding value: as long as technology progresses that is.

Let's go right to the limit in a thought experiment though, shall we? Imagine the Wishes Fairy was available to us and we could wish for the wet dream of the Green economy. All energy would come from renewable sources, there would be no new appropriation of physical resources from the environment. We recycle everything, which given enough energy we can do, and we make sure that none of the renewable systems, like forests or fresh water, soil fertility, have too much strain put onto them. We tread lightly upon this earth perhaps.

Could we, in this situation, still have economic growth? Yes, most certainly we could. In fact, I'd say that it's absolutely certain that we would have economic growth: for we are really very inquisitive monkeys when all is said and done. That revolutionary new turnip-hoeing technique which allows an extra hour of leisure in a day, time spent taking singing lessons, is economic growth. That slicing and dicing of the silicon wafers so that more can be made from the same ingot is economic growth. If, even in a system where there is no new resource use at all we will still have economic growth, then resources themselves cannot be the limitation upon economic growth. A limitation, most certainly, but not the one which makes the entire concept simply not possible.

Another way of saying much the same thing is that there is good economic growth and bad economic growth. Good and bad being defined by those who worry about resources. There is economic growth that uses more resources, this is certainly true. Pumping up more oil increases GDP. Getting more labour out of the house and into the market does so: mining more iron ore, baking more cement, all of these contribute to economic growth as conventionally defined. Yet there is also economic growth which comes from creating greater value while using the same resources, fewer resources and possibly even no resources at all. By being more efficient in our use of those resources, a point which we will see is important when we come to discuss markets.

To make the same point in a trivial manner: sending a virtual red rose on Facebook is economic growth. Those who send them pay to do so, so yes, this does turn up in the GDP figures. Those who receive them presumably enjoy doing so, so we've an addition to human happiness: if those who sent them get the legover that we might assume prompted the sending they too will be happier. As long as people keep thinking up new ways to send these virtual red roses (and given the human propensity to think up new and interesting ways to get sex we might assume that there will be) then we can continue to have economic growth.

Our bottom line here is that the physical world is not the defining limit upon economic growth: human ingenuity is. For what we are measuring as economic growth is not the consumption of the resources provided by that physical world but the value that human beings place upon their own consumption, that consumption having no necessary relationship to resources at all.

Notes and References

1. http://www.grida.no/publications/other/ipcc_sr/?src=/climate/ipcc/emission/
2. http://www.neweconomics.org/sites/neweconomics.org/files/Growth_Isnt_Possible.pdf

4   Globalization

By means of glasses, hotbeds, and hot walls, very good grapes can be raised in Scotland, and very good wine too can be made of them at about thirty times the expence for which at least equally good can be brought from foreign countries. Would it be a reasonable law to prohibit the importation of all foreign wines merely to encourage the making of claret and burgundy in Scotland? But if there would be a manifest absurdity in turning towards any employment thirty times more of the capital and industry of the country than would be necessary to purchase from foreign countries an equal quantity of the commodities wanted, there must be an absurdity, though not altogether so glaring, yet exactly of the same kind, in turning towards any such employment a thirtieth, or even a three-hundredth part more of either.

Adam Smith, *Wealth of Nations*, Book IV, 2.15.

Apologies for the outbreak of 18th century prose there: but it is mandatory to quote Adam Smith in any book about economics. Failure to do so leads to tutting or worse, tsking.

The point that Smith is making here is obvious: why make something expensively yourself if you can get it cheaper elsewhere? We can also see that the use of glasses, hotbeds and so on is a use of more resources than are used in simply letting the grapes ripen in the French sun: whether you end up with Scots or French verrucas in your wine will obviously be a matter of taste but it's obvious that fewer resources would be used in the production of the wine in France.

The resources used in transport could be higher than those used in local manufacture, this is true, and we can measure that by what we're charged for the transport. Or when considering things external to market prices, those pesky externalities we've already mentioned, we could do the direct calculation. That there's very little international trade in potatoes while a very large one in wheat (to the point that much Italian pasta is made from Canadian wheat) does tell us something about transport costs in relation to the value of the commodity being transported. Or even that much wheat for bread consumed in Britain is from Canada, while much British grown wheat goes abroad to feed animals.

This is a combination of Smith's point about trade with the one made by David Ricardo so it's worth untangling the two as they're so often misunderstood. As well as the wine example Smith pointed out that:

> *What is prudence in the conduct of every private family, can scarce be folly in that of a great kingdom. If a foreign country can supply us with a commodity cheaper than we ourselves can make it, better buy it of them with some part of the produce of our own industry, employed in a way in which we have some advantage. The general industry of the country, being always in proportion to the capital which employs it, will not thereby be diminished … but only left to find out the way in which it can be employed with the greatest advantage.*

(From Book IV 2 again)

In short, if other people can make it cheaper than we can let's get it from them and turn our efforts to something else. Part of this comes from the obvious matter of natural endowments: not even the most insanely anti-trade of environmentalists thinks that we should be growing our own pineapples in Britain. Having just one such tree for show in Kew Gardens is quite enough to show that we don't have the resources in glasses and hothouses to feed the nation's gammon and pineapple needs from local resources. But Smith went further in his discussion of why trade makes us all so much wealthier.

The key to trade is the division of labour, the specialization of labour and then the trade of the resulting production. Smith's example of this division and specialization was a story about a pin factory that has been boring economics students ever since. The basic idea is so trivially simple that it's rooted in aphorisms in nearly every language. Jack of all trades, master of none might be English but similar phrases occur in most other languages. What is meant of course is that if we concentrate on one task, one procedure, then we'll be better at it than if we try to do everything. If every one else similarly specializes then collectively we'll get more and better things which we can then swap. Yes, we can go on and on about this, as Smith and hundreds since did, but the essential concept is obvious. Whether we're saying that you grow the apples and I'll make the cider and we'll get drunk together or that I'll write a book and you'll read it and we can have huge fun shouting at each other about it afterwards, we're still saying let's divide the labour, specialize and trade the results: a process that makes us richer.

Ricardo's addition to this, comparative advantage, is not quite so trivial, as Paul Samuelson pointed out:

> Nobel laureate Paul Samuelson (1969) was once challenged by the mathematician Stanislaw Ulam to "name me one proposition in all of the social sciences which is both true and non-trivial." It was several years later than he thought of the correct response: comparative advantage. "That it is logically true need

not be argued before a mathematician; that it is not trivial is attested by the thousands of important and intelligent men who have never been able to grasp the doctrine for themselves or to believe it after it was explained to them."[1]

The result of this idea is that it is never possible for there not to be gains from the specialization of labour and the resultant trade. It's also, as Samuelson points out, something which is grossly and widely misunderstood.

It is that it doesn't matter how good or bad I am at anything compared to how good you are at any and every thing. That isn't the comparison we are thinking should be made. The comparison is between how good I am at doing things and how good I am at doing other things. The specialization I choose, the work that is divided to me, should be what it is that I am least bad at.

Think of the extended family barbecue: there will be the idiot cousin there. No, really, every family has one (and as with poker, if you don't know who the mark is at the table then it's you) but do we, when preparing the garden feast, insist that said idiot do nothing? Everyone else there is better at everything: including the patting of mince into hamburger patties. So should Fred (no, I don't have a Cousin Fred) be told to do nothing, not even shape patties? No, that isn't usually what we do, Fred does the thing which he is least bad at, even though he's worse at everything than everyone else, and pats shredded cow into shape. We thus get meat that goes into buns as well as all of the other necessities, the coleslaw, the ribs, the Pimms and the cucumber sandwiches for later (for we're certainly not going to let Fred near either flames or sharp edges, not after that barbecuing the cat on the embers of the gazebo last year). Fred has added to the wealth, the production, through this specialization and division of labour, as Smith said he would, and done so by doing what he is least bad at as Ricardo pointed out he should.

Our wheat example above can now be explained: sure, it seems a bit wasteful that the UK both imports and exports wheat but they're actually different types of wheat. Our natural advantage is in soft wheats suitable for animal feeds: that of Canada in hard wheats suitable for pasta and bread.

This possibility of using fewer resources to produce via trade means that in any economic system we're going to have a higher standard of living for a given level of resource use if we trade: alternatively, that we'll have a lower level of resource use for the same standard of living as we could get through non-trading. Trade thus lowers resource use.

Yes, I know, that's not the way it's normally portrayed. Fortunately the fuss over food miles is fading as people have been pointing out that 50% of the emissions produced by the food on your plate comes from bringing it back from the supermarket: nothing to do with whether it came from the next village over or from New Zealand. There is still the possibility though that those externalities make this untrue. In our case, here, talking about climate change as we are, the CO_2 emissions that come from transport.

Fortunately, there have been a brace of decentish studies looking into precisely this point. One, coming as it does from the New Zealand lamb exporters' trade body, might be seen as tainted at source, but their figures showing that raising a lamb there and shipping frozen to the UK has lower emissions than raising one on a Welsh hillside seem fair. They don't need to use fertiliser, nor warm up the lambing sheds, nor keep the sheep indoors in the snowstorms. The other was done by our own DEFRA (or whatever they've changed the name to now) and looked at tomatoes. Those grown in Spain, under the sun, and trucked up have lower emissions than those grown in glass houses (well, heated polytunnels now) in the UK. We're actually reducing our emissions by importing our food.

Sometimes we're reducing our emissions by importing our food that is. Those emissions being external to market prices means that we can't use our simplest measure of what is in fact using more resources, the price system. This is indeed an important point and we'll see how we should deal with that later when we get to carbon taxes and cap-and-trade, both ways of getting those externalities reflected in prices.

The most important point to make about both Smith and Ricardo on trade though is not this simplistic exposition of their views, as above. Anyone who has ever seen a home-made bookcase will be convinced of the value of the division and specialization of labour. No, the important point is that there's no particular limit to either of these ideas. The division and specialization of labour, the resultant trade, us all doing what we're least bad at and swapping the resultant production, there's no reason at all that this should be limited to exchanges with idiot cousins. There's no reason to limit it to households, families, villages, towns, counties or countries. Indeed, there are very good reasons not to so limit it: there are few villages or towns that could ever support a computer manufacturing business: not even countries can do that.

Globalization, the free movement of goods around the world, free of tariffs and restrictions, in fact makes us able to divide labour with, specialize in labour beside, trade with, all 7 billion of our fellows. To our great benefit as above.

There are those who argue with this: Karl Polanyi was a mid last century economist who was insistent that trade with someone you knew was more valuable than trade with someone you didn't. Trade should be, in his view, something which was a mutual and reciprocating web of obligations, not something intermediated by the use of money, filthy lucre. Well, yes, but that would make my own working life rather difficult. Yours too in fact.

My day job involves the wholesaling of a specific metal into a specific industry. Tiny amounts of something most have never heard of but the trade covers at least six continents. Even the production of the raw material involves three: and the final product goes back to all six inhabited ones (and I'm told to the airbase at the south Pole as well, making the set of all the continents). The metal itself is vital for making street lights with: it's what makes the bulbs themselves work as well as they do. It's a tiny industry, it has the sort of turnover which would support, ooooh, perhaps one person in the world? That's all it does support, certainly, the trade in this metal (as opposed to the much larger industry which uses it) supports, to my certain knowledge, myself and myself alone. Something which means that all 7 billion of you need not worry about how to make the light bulbs that mark your totter back from the pub at night. The world needed a specialist in this subject and in this flavour of this universe it got one in the form of your humble author.

Caroline Lucas once gave one of those 90-second little speeches which are what the European Union has reduced democracy to in which she said that she wanted global cooperation, not this awwwful global competition. Not realizing that trade is exactly the way in which people cooperate. She then went on to say that of course we needed protection against certain foreign products: meaning that while she wanted global cooperation she didn't want to cooperate with Johnny Foreigner. A slightly odd position really. For as the above shows, what global trade allows is that I cooperate with everyone on the planet (no, really, it is near everyone, we've supplied about 80% of global consumption by the lighting industry for the past 15 years) when they stand under a streetlight. Every drunk looking for keys, every tart advertising by leaning on a lampost, every dog using the light to aim, every yummy mummy taking the kids back from late night music lessons has been cooperating with a middle-aged man in Portugal.

I have to say that I cannot really see the problem with this global trade stuff.

However, it's possible that some will not be convinced by economic theory (economists are hardly flavour of the month right now) nor my talking my own book. So why don't we go and have a look at the scientific consensus? The scientific consensus on climate change, that is, the IPCC itself?

What appears to be little understood, and something of a teethgnasher for economists, is that the entire structure of the IPCC is based upon economics. It's not exactly rare to hear someone spouting off about how climate change means we need an entirely new economic order: maybe we do and maybe we don't but it's a remarkable conclusion to come to from a report based upon the most basic of entirely mainstream economics. The use of mathematics to prove that there really are only 180 degrees in a triangle would not be taken as proving that we need a new mathematics in which there could be 200 degrees in a triangle, now would it?

That economics comes in the already mentioned Special Report on Emissions Scenarios[2], or SRES. Think through the entire process of the IPCC for a moment: we've got as our output a number of guesses (yes, they are guesses. They're reasonably well-informed guesses, but they are still guesses. This absolutely isn't like the physics of knowing where the Moon will be in 35 years) about what the climate will be like in 80 and 90 years' time. There are a few fairly hairy assumptions in there about things like feedbacks: things we're very much not sure about as yet. These guesses have to be based upon something and indeed they are. A combination of suppositions about what certain concentrations of greenhouse gasses will do to the climate and how many greenhouse gasses there will be to influence that climate.

The influence of the gasses comes from a number of different climate models: those pieces of software that various groups are running around the world. The amount of those gasses, the numbers which are put into those models to see what the effects will be, comes from economic projections. They have to come from economic projections, obviously. For our three determinants of how many gasses there will be are how many people will there be, how rich will they be and which technology will they be using? Two billion poor people using non carbon energy generation systems will be emitting different amounts than 16 billion richer people stocking up the coal fires: Obviously.

Those *au fait* with matters environmental will see that this is a version of Paul Ehrlich's famous IPAT equation. Impact (the environmental impact that is) will equal Population times Affluence times Technology. Of course, in order to use one of Paul Ehrlich's ideas we actually have to correct it before it is umm, correct. The impact of a population is not multiplied by the technology it uses, it is moderated by it. Higher technology societies tend to have lower impact upon their environments than lower technology ones: "T" moderates, not multiplies "I".

Just in case this is frying the odd neuron in those who think that Ehrlich has ever been right about anything, consider the current state of the planet. Certainly not perfect, there are far too many suffering poor, forests are being cut down rather too quickly and sadly we have entirely idiot methods of managing fisheries still (yes, Common Fisheries Policy, we are lookin' at you) and we've this nagging worry over climate change, but let's look at the technology in "T".

The lowest form of human technology is the hunter-gatherer lifestyle: it predates the existence of the human species by some millions of years after all. If there were 7 billion people on the planet trying to live such a hunter-gatherer lifestyle, well, there wouldn't be 7 billion people really. There would be bare rock, empty oceans and the scattered skeletons of those 7 billion (only the last to die of which would not have human tooth marks on) as there simply isn't enough land, not enough primary production by purely natural means of calories to support 7 billion people. So we can without fear or favour say that the impact of 7 billion humans with ground zero technology would be pretty high.

We now have 7 billion people with an average wealth (GDP per capita rather) of around $8,000 a year and while we'll all admit that there are environmental problems our impact upon the environment really isn't as final and terminal as using that ground zero technology. Which means that as we've the same number of people who are richer but the impact is less, then, well, basic algebra tells us that higher technology must moderate, not increase, the effects of wealth and population. So we can happily assume that higher technology reduces impact: and this is indeed what the SRES assumes when looking at what the future will bring. A higher technology world is a cleaner one, one where we tread more lightly upon the Earth as, as the good little environmentalists that we all are, we would wish. Technology is our friend in short, not our enemy.

In more detail the SRES assumes that there are four major ways the world could develop: we can have more or less globalization and we can have more capitalism red in tooth and claw or we can have a more caring, sharing, worry-about-the-distribution, economy: what I tend to term as the "Kumbaya" economy. We'll all do what no human society as yet has done and share everything among all delightfully and sing that blasted song around the campfire when we have done.

This gives us four families: A1 is global capitalism, A2 is a non-global, localized and regionalized capitalism. B1 is a globalized caring and sharing and B2 is a localized and regionalized kumbayaism. Within these four families we then have 40 what are called "scenarios". For the families we set the population, the level of wealth and the general economic structure: the scenarios then add the specific technological paths that such societies might follow. As you can see, this gives us the "I", the impact of the IPAT equation which allows us to go off and plug these results into the climate change models.

We could, if we wished, regard our choices about which way we go, local or global, capitalist or not, as simply moral choices. There are those who say that capitalism is immoral because man exploits man while not capitalism is moral because everything is the other way around. Similarly there are those who feel, and it may actually be true for they themselves, that trading only locally, only with those they know, have a web of interactions with, makes them better off: richer in fact. But here we're concerned solely with climate change. We're not, after all, trying to work out how to make the world perfect: we're just interested in stopping the oceans from boiling Flipper or rising to drown us all.

When we look at the outcomes from these various musings upon how the world might turn out we find something which is entirely at odds with what we're all being told. The results of a globalized world, in terms of climate change, are better than the results of a localized or regionalized world. A1 can produce less emissions than A2 and B1 definitely produces fewer emissions than B2. In fact, the two outcomes which produce the lowest emissions are B1 and A1T (a scenario in the A1 family which assumes rapid technological advance in non-carbon forms of energy generation).

That is, in terms of climate change, global beats local.

This isn't, I know, what we've been hearing over recent years. The incessant (strident even) insistences that we must become self-sufficient, grow more of our own food, lower our trade with those far away, reverse the globalization of the world: all of these are shown to be wrong by this, the scientific consensus upon climate change.

Whether you want a more or less capitalist world is, at least at this point, something which we can leave to your moral precepts. But as we're all bound to follow the science these days we cannot leave the issue of globalization or not globalization to your (or even my) prejudices. The science is clear: more globalization will reduce the impact of us humans upon the climate. As that's what we're actually desiring, the reduction of our impact upon the climate, then our necessary, our just and righteous, course of action is to go gung-ho for globalization as fast as we can.

That is, after all, what the scientific consensus on climate change tells us.

Usefully, it's also what Smith and Ricardo have pointed to as well: for any given level of resource use we will be richer through trade. For any given level of wealth we will use fewer resources through trade: our impact for any given level of population or affluence will be moderated through using the technology of globalization.

Isn't that an interesting finding for our modern age: Dead White European Males do have something interesting to say about our modern problems?

Notes and References
1. http://www.wto.org/english/res_e/reser_e/cadv_e.htm
2. http://www.grida.no/publications/other/ipcc_sr/?src=/climate/ipcc/emission/

5 Population

People eating, people washing, people sleeping. People
visiting, arguing, and screaming. People thrusting their hands
through the taxi window, begging. People defecating and
urinating. People clinging to buses. People herding animals.
People, people, people, people.

Paul Ehrlich, *The Population Bomb*

Just as Adam Smith quotes are essential in a book about any
form of economics so are Paul Ehrlich ones in one about any
matter environmental: you can tell which sort of environmental
book you're reading by how such pronouncements are treated. A
reverential reference suitable for a distinguished professor means
you're reading a book by a committed environmentalist; a
rhetorical parp! in his direction by a rational environmentalist.
We here are of course parping (just imagine how tortured the
prose would have got if they were defecating into the taxi!
Fortunately this was Indians, not Mark Oaten) for the
distinguished professor has spent the 50-odd years since he
penned those sentences ignoring everything we know about
what reduces population: Wealth.

We'd all agree that there is in fact some limit to the number
of human beings that can be usefully (or even possibly) fed and
watered upon this one and only planet that we've currently
managed to lay claim to. What that limit is depends rather upon
what we want to use as the limiting factor: we can be absurd and

use the weight of the average person as compared to the weight of the Earth if we like. We can't have more people than there is planet for them to be made from, for example. Humans also create warmth purely in the process of living (this is the principle that an igloo works on, the heat generated by a body itself) and we might say that the limit to the number of humans is the heat they give off. Too many and we'll be creating more heat than can escape from the atmosphere and thus we all boil. Which is rather where we came in with climate change really, although here it's our activities rather than ourselves causing the potential problems.

Somewhere below that is the limit on how many people we can have, well fed, fat and happy, at whatever level of technology we happen to have available to keep everyone well fed, fat and happy. That, of course, is where all the shouting matches start. At one end we have the Optimum Population Trust who think the UK can only support 17 million and at the other you have people like myself who think that we're never going to get anywhere near any of the physical limits. Because of that aforementioned wealth in fact rather than the liberal dispensing of contraceptives which seems to be the usual cure for over population worries.

But let's leave aside what actually is the population limit, assume that everyone is indeed worried about it and then ponder instead on what we might do about it. The usual suggestion is that we should simply pump out contraceptives to everyone and then they'll stop having so many damn anklebiters. Something which on the face of it sounds a little odd for, oddly to one such as myself, many people seem to rather like having the little snot machines, even go out of their way to do so. It isn't just that people like the fun of sex and hope to dodge the stork: people actively desire children, plan their lives around the acquiring of them.

We could also note that family size decreased some years before there were effective methods of either barrier or chemical contraception showing that it isn't the absence of contraception that prevents control of family size either. That Catholic women are urged to get rhythm (no, Carl Perkins was singing about something else) shows that, while not perfect, there are other workable methods. Anyone who has managed to get past page four of *The Joy of Sex* (or has been in a relationship for more than a few weeks) knows that there are plenty of ways of enjoying the sex bit while making sure that conception is not possible.

We could even, if we were going to be all scientific about it, look at what the science says about fertility[1]:

> Ninety percent of the differences across countries in total fertility rates are accounted for solely by differences in women's reported desired fertility.

Yes, that paper does go on to point out that the availability of contraception only explains 10% of the changes in total fertility: quite a number of couples out there must have worked out that beat thing or been in relationships for more than just a few weeks.

So if we want to change total fertility, the first stage in getting to a slow down and possibly a reversal in population growth, it isn't enough for us simply to make barrier or chemical contraceptives available to all who would ask for them. There aren't enough who would like to but don't currently use them to make the difference we desire. What we would have to change is the number of children that women desire in the first place. This, of course, doesn't mean that we shouldn't go right ahead and supply them to those who would like to use them. Just that we shouldn't believe that this is going to change behaviour all that much, solve our perceived problem.

So what we'd really like to find out is what is it that changes the number of children that women desire? We could, if we wanted to be trite, go back and look at our economic models from the IPCC and note that the two globalized families have population peaking around mid-this century then falling to 7 billion. The two non-globalized families have population at century's end of 16 and 10 billion, for A2 and B2 respectively. Clearly, globalization itself reduces the number of squealing brats any one woman desires to pop into the world and should be supported for that reason alone. But that would be trite: taking as the ultimate cause what is in fact the proximate one.

For what the globalized models produce is vastly greater wealth per person than the non-globalized ones. Average GDP per capita in 2100 is, for A1, $75,000, A2, $16,000, B1 $50,000 and B2 $25,000. Our scientific consensus is showing us that there is a definite correlation between higher incomes and lower population: it's not that ever higher wealth produces ever fewer people, rather that over a certain level population begins to fall rather than continue rising. The earlier we get to that level the lower the peak of population will be and the sooner it will begin falling. That A2 model, by the way, still hasn't reached it at that income level.

Our question now therefore is: what is the mechanism by which higher incomes produce lower desired (and thus total) fertility? After all, we would expect that richer people have more resources and thus are able to raise more children. That's certainly how it has been through human history, the rich have more children who survive and thus have more grandchildren. What is it that short circuits this when we get to some general level of high enough incomes? Why does wealth lower the general birth rate even as it has, historically, led to increased populations?

To answer this we need to delve into the real meaning of life. You can take it as the Biblical "go forth and multiply" if you like, although I would prefer the more Darwinian "having grandchildren is winning". The aim and purpose of our whole stay in this vale of tears is the perpetuation of our own genes. This doesn't have to happen through the propagation of our own directly of course, it can happen through aiding the success of the family, the group (as JBS Haldane once said, "I would lay down my life for two brothers or eight cousins"), but the general name of the game is to make sure that we have enough children to ensure that some of them survive to go on and have grandchildren. Given average lifespans that's about as far as we've historically been able to take it, the bit that we can actually observe happening.

It can be difficult for us in this antibiotic'd, NHS'd age, to realize quite how fragile were children's lives only two or three generations ago. Or how fragile they are in some parts of the world today. It's not unusual to see numbers like one in five fail to survive their first year, one in four their first five years. A visit to an older graveyard will show any number of infant graves: try counting them sometime, you would, if such happened now, think that a plague had passed through. But this was simply (while a tragedy each and every one, that weighed just as heavily upon the minds of our forebears, or of those stuck in the child killing misery of the Third World now, as the loss of a child would to us now) the result of bad water, bad sanitation, the lack of vaccines: the things which wealth and technological progress can now provide.

Mothers' lives were also a great deal more fragile than they are now. Current maternal death rates in advanced countries are around 10 in 100,000: for every 100,000 live births ten of the mothers die. Out there in the wilds it can be as high as 2,000 and it's thought that the historical average over our time as a species was around 1,000 per 100,000 (yes, it is possible for bad practices, such as bad health treatment, to make matters worse than nipping behind a hedge to give birth. The first maternity hospitals are now notorious for raising the death rate from puerperal fever via cross infection to as much as 40%, until Ignaz

Semmelweiss managed to convince doctors to wash and change between post mortems of dead mothers and delivering the next child).

I've no doubt that you can differ with me on this but I take it as a moral precept that we should be aiding those currently suffering such historically normal but currently obscene child and maternal mortality rates in lowering them. But there's a wrinkle to this: more children surviving, more mothers surviving childbirth, won't that lead to the very population increase we're hoping to avoid? No, actually, it won't, as it didn't for our own societies, and that's where the economics comes in.

We are generally urged to look at a number of correlations when we are similarly urged to think about how to reduce birth rates. We can, for example, see very well that educated women tend to have fewer children than the uneducated (we see this here in the UK just as much as anywhere else), so if we educate lots of women then there will be fewer children. Similarly, women who know about contraception, indeed about sex, tend to have fewer children. Women who work in the market (rather than solely household labour) tend to have fewer children so why not get more women working outside the home? Women who have greater social power tend to have fewer children: certainly more gender equitable societies have lower birth rates.

All of these (and more) are fine as correlations, but they don't get to the ultimate cause of the fall in desired fertility which is what leads to the fall in total or actual fertility. It isn't education itself that makes a woman desire fewer children: it's that the ability for women to have an education is correlated with other things which reduce desired fertility. That thing they're both correlated with, as so are women working in the marketplace, so are falling maternal mortality rates, falling child mortality rates, is wealth. Let's promote gender equity to reduce the birth rate, why not? Except greater gender equity comes largely from moving from a peasant society, dependent upon male musculature, to a richer one which uses machines for that, allowing greater economic and social equality between the sexes. The economic capacity of that nation, country, to provide all of these highly desirable things is wealth and it's wealth which enables their provision, their existence. And of course they reinforce each other: fewer children being born will lead to more women being able to work for more years, meaning further economic growth.

Think of a, by current standards, by current western standards, a very poor society. Something like a quarter of children born never reach their fifth birthday. There is always also the possibility that even those who do will die in their youth, before they themselves have a chance to breed and produce those desired grandchildren. Each and every pregnancy leads to a one-in-50 chance that the woman herself will die. Desired fertility, the desired number of babies that will give a reasonable chance of leading to the desired grandchildren, might be up at 6, or 7, 8 perhaps, children. Don't forget, if you gamble and have fewer children and that famine, that epidemic, that war, does pass through, then you've lost. Lost everything that 4 billion years of evolution have been striving for. And there are parts of the world where this is, or at least recently has been, actual fertility. But having this number of children leads to a 10%, 15% likelihood (it's not linear as the greatest risk is with the first child) of the death of the mother in childbirth. Not to mention at least two decades, the majority of the average adult lifespan in such a poor society, spent caring for newborns and infants.

Which is where the rather harsh calculus of economics comes in. In such a society, is investment in the education of women actually rational? They are already fully occupied in that most vital of tasks, creating the path to grandchildren.

Why educate them, why invest in them when there is nothing more important that they could be doing? If most of adult life is to be spent suckling and weaning, with a very good chance of an early death in the process, why spend money or resources, things in very short supply ('coz this is a poor society, see, that's what a poor society means), on someone who has a very good chance of not using it and thus wasting the invested resources?

It is only with the creation of a little bit of wealth that this, admittedly extremely harsh, calculus changes. Wealth in its true sense of course, not just moolah in the hands of a man who might spend it down at the pub. Wealth in the sense of children living, of potable water to prevent the diarrhoea that kills millions each year, bednets and DDT to control malaria, wealth in the sense of a health care system: any health care system in fact. Only when a society is able to provide these things and yes, that does mean that the society has to become wealthier in the more traditional meaning of having more money, having an economic surplus, does it then make sense for there to be investments in the education of women.

At the same time, a society that is rich enough to invest in such education is also going to be a society which is moving away from pure human grunt as its power source and thus will have at least the possibility of openings for women in its economy. The education of women is now not only possible it is also valuable, capable of providing a return. Things which are both possible and valuable tend to be things which get done: yes, it really is true that human beings are greedy and they will indeed do things which provide them with a return just as they tend to not do things which don't.

Once this virtuous spiral has started then it tends to continue. It has done everywhere it's happened so far at least, although whether it's going to survive a meeting with an absurdly patriarchal religious society like some of the ones out there is currently unknown. There is hope of course for even those places we think absurd in their gender segregations and oppressions today are not notably worse than was our own society only a century or two ago.

There is also the economist's favourite point to mention: opportunity cost. A richer society is one in which people have more choices. To work or not to work, to become a career woman rather than a peasant in the paddy fields, to study literature rather than simply exist. The more choices there are the less we tend to do of any single one of them: we do have those who bear 8 children in our own society but with all of the opportunities that we have most decide on fewer so as to have the time for more of the alternatives.

The result of all of this foray into one of the more dismal parts of the dismal science is that we find that we don't just desire economic growth, we require it. We desire it for the simple reason that it will make lives as they are currently lived better. But we require it because it is also the only thing which will lead to a reduction in population growth and then to a reduction in total population. Plus, for those who are interested in such things, an end to the holocaust of dead babies and mothers whose corpses litter graveyards around the world. Holocaust is the correct word by the way, even if not The Holocaust for the numbers of deaths per year from poverty are higher than the number per year from Hitler's monstrosity.

Again it seems reprehensible, foul even, to calculate the rights and desires of women and children, their very lives, purely in terms of economic growth, that increase in the value of goods and services produced each year. But it is indeed true that population growth will stutter and then go into reverse only when women have more productive things to do with their lives than pump out ever more babies in an attempt to reach that goal of human life, to multiply, to have those grandchildren. They will only have those more productive things to do, only be invested in so that they are productive, when there is the wealth to cut those child and maternal mortality rates and for that, that reason alone, we have a moral duty as I see it to foster and encourage economic growth.

Notes and References
1.   http://ideas.repec.org/p/wbk/wbrwps/1273.html

6   Markets

> *Nordhaus brackets the growth of real wages over the past century as somewhere between a 20-fold and 100-fold increase. Alan Greenspan ... has suggested adjustments of the statistics that lead to an estimate of a thirty-fold increase of material wealth over the past century.*
>
> Brad Delong

That everyone got richer in the 20th century isn't quite true: but that everyone who took part in the 20th century did, is. As long as you'll allow me the conceit that taking part in the 20th century means living in one of the industrialized countries. It isn't true of certain parts of Africa, where there are certainly places poorer now than they were 50 years ago, 60, under colonial rule, and obviously it isn't true of those few people still living the hunter-gatherer lifestyle in the depths of this or that forest. While there are those who disagree, I take it to be axiomatic that we'd like all to enjoy the same good fortune that we do, living richer and longer lives than our forbears did. The interesting question though is how we might bring that about?

As we saw earlier, further economic growth is possible, and desirable for population reasons if no other. But we've also seen that there are two types of growth. Growth which comes from the deployment or use of more resources and growth which comes from adding more value to the resources at hand. It's a common trope that we cannot have more of the using resources kind of growth as there simply aren't enough resources to be used. While we might argue with this on sophisticated grounds, let's simply agree at this point. OK, you're right, greenies, we really do have resource limits and they're sufficiently tight that we cannot simply keep digging up more of the planet to make everyone more shiny gewgaws. Those who point to the fact that coincident with that growth of the last century there was also a growth in resource use and consumption are, for the sake of our argument here at least, to be regarded as correct. Well, almost correct, for while to quote one Nobel Laureate quoting another might be overdoing the arguing from authority point, there is still this from Paul Krugman[1]:

> How, then, have today's advanced nations been able to achieve sustained growth in per capita income over the past 150 years? The answer is that technological advances have lead to a continual increase in total factor productivity – a continual rise in national income for each unit of input. In a famous estimate, MIT Professor Robert Solow concluded that technological progress has accounted for 80 percent of the long-term rise in U.S. per capita income.

So, there are two types of economic growth and we cannot have one, so how can we make sure that we have the other? How might we try to swing things so that we can have that 80% of last century's economic growth, the one that comes from the adding of more value (which is what technological advance is) rather than the increased consumption of inputs?

The go-to economist on this point is one William Baumol (who, along with the other William, Nordhaus up above, is on everyone's perennial lists of those who will get the Nobel if they live long enough and no more youthful tyros like Krugman are allowed to queue jump). His point in this argument is that there's a very large difference between invention and innovation: but we must be careful because these are words that are, in different arguments, defined differently. In much writing about business, invention is thought of as the creation of some really cool new thing. Completely different from what went before, the sort of thing that has reviewers throwing around bits like "creating an entirely new mould" and, appallingly, using the word "paradigm". Think the invention of computers for example, or the personal computer, or an entirely new way of building a vacuum cleaner as James Dyson did. Innovation is thought of as the incremental changes, the polishing up of one of these originally path-breaking ideas, as we go through the iterations of model years and small changes to manufacturing efforts.

In these senses invention is usually thought of as coming from the outsiders, the mad inventors in their sheds (Dyson did have a shed if not the madness, that lack of madness being proved by his purchase with his hard earned cash of a quite beautiful estate with Capability Brown gardens just north of Bath, an estate I'd long had my eye on other than for that quite gross unfairness of this universe that I was short some £34,999,999.96 of the necessary price. That the house was originally built on the proceeds of running the island of Barbuda as a stud farm for slaves worried me not: *paradisus non olet* after all and the last of the Codringtons was a nice old boy when encountered in the pub.), while innovation comes from the large companies. Part of this is the rueful admission that large companies really aren't all that interested in breaking the mould: they've far too much already invested in the current way of doing things to want to do any more than repaint the applecart rather than overturn it. That IBM brought us the PC as we know it would seem to militate against this idea (although it is a generality, not a hard and fast rule which can be refuted by one example), but the fact is that the IBM PC was created by a small maverick group, and it almost certainly would not have been launched if the larger company had understood it. That IBM was nearly bankrupted by the changes in the computer market that its own PC created is further an object lesson in why large companies are often not all that keen on that new paradigm stuff, preferring the incremental changes that can be controlled – the more usual business practice of bringing out an upgrade to the AS 400 line, a slightly faster processor and a bit more memory, each year, for example.

Baumol uses the two words in rather different senses (no, this isn't just redefining the language so as to make his theories work, he's not a politician). It's just that he uses invention and innovation to mean slightly different things. Invention is that creation of the really cool new thing. Innovation is the adoption and use of it by the wider society. It's with this definition in mind that we can begin to look at how our economy needs to work if we're to have economic growth within our resource constraints.

As Baumol has pointed out, the Soviets created some pretty spiffy stuff in their laboratories, so we can't say that a planned system is incapable of invention. I would second that of course, as someone who has spent part of the last decade marketing one of those spiffy Soviet inventions – a new and wondrous aluminium alloy. We could also note that lots and lots of the splendid new things we get in our own society also come from a standing start in the state/university system. Invention, in short, can happen in pretty much any economic or political structure. But what matters is not that such things are invented, it is that such things are used: it is only by using the technological advances that we get the value added which is our definition of that economic growth. Here there really does seem to be a difference in economic and political systems: some allow such use to spread quickly, others do not. State, or planned, economic systems seem to deter the swift adoption of such new spiffiness whereas market systems promote it. Given that our desired economic growth, the rate of it, is determined by how fast spiffiness spreads through the economy we will therefore want to have a market-based economy.

Let's take one of the modern world examples of what is being talked about here: mobile or cell phones. Quite a bit of research has been done on how the adoption of mobile phones aids in boosting economic growth. One finding was that an increase of 10 people per hundred who had a mobile boosted the growth rate by 0.5%. No, that's not 0.5% of the previous growth rate of, say, 3% (ie, 0.15% of GDP), that's 0.5% more of GDP: an astonishingly high rate for a single technology. It should be said that these growth rates were found only in countries where there was not already a decent system of telecommunications. This is the addition to growth of having telecoms, not mobiles specifically. Thus this growth effect has been observed in those Third World countries that do not have a land-line network (and, to be obvious about it, now never will have as a mobile network is vastly cheaper to deploy than a land-line one). A similar effect can be seen in economic history from the deployment of land-line networks where those did and do exist. Being able to communicate, being able to find out what things are like 5 miles down the road without having to go there, does indeed produce economic growth.

Why this should be so can be seen in several studies that have been done about sardine fishermen off the coast of Kerala, in India. It's a pretty low-tech operation, a rickety boat, an ancient diesel engine, a few hungry, dhoti-clad men and a net. Push off in the morning and see where you can find the fish, catch what you can and bring them back to shore. Two way radios are an expensive luxury in such an environment and so, before the arrival of the mobile phone networks (yes, they do work far enough out to sea to cover the usual fishing grounds) there was no easy manner for the fishermen to work out who would be the best buyer for their catch.

They would simply head back to port and hope that the buyers hadn't already been stocked up by earlier, luckier, fishermen, or simply gone home empty-handed. There's also the possibility of drying some of the fish (for a lower price than fresh, but which gives the ability to sell not today but tomorrow), of shifting between different ports where different buyers might be under- rather than over-whelmed and even, when finding a shoal that is too large to be captured by one boat alone, of calling over a few friends on other boats and saving them some diesel in their search patterns. All of these things are impossible to do without a communications system, of course. Mobiles became that desired communications system and allowed all of the above things to happen: switch port to one where there are still eager buyers, call a buyer and tell him to wait as you're on your way in. Call Cousin Anish and tell him to get over here pronto – there's more sardines than you can shake a stick at. Even, sometimes, the bad news that all buyers are entirely stuffed with all the sardines they want today: in which case, save the fuel and come on home.

The net (sorry) effect of all of this telecoms network (yes, sorry) has been that sardine fishing has become more profitable. More than just that, the costs (of fuel etc) to the fishermen have gone down, the cost of sardines to the consumers has gone down, fewer fish are wasted by not finding a buyer and, as far as anyone can tell, there's still the same number of sardines out there in the ocean. This is pure economic profit: simply the addition of a method of communication has made people richer, both the fishermen and the consumers, by making one little part of the economy more efficient. It's not the invention of mobile phones that has done this (and so we can sidestep the entire whining match about whether Swedish government subsidies to Ericsson were where it all started or not) it's the innovation, the finding of ways to use them which has done.

Now we might think that a planner could see the value in such communication and that such a planner would therefore equip fishermen with the technology once it had left the laboratory. Which really isn't quite the point at all: what we want to look at rather is which system, planning, those good and wonderful, omniscient, public-spirited, bureaucrats organising the economy on our behalf, or the chaos and anarchy of a market-based system, which of them produce more phones in the hands of more people who are able to increase economic efficiency howsoever they might find a way to do so?

There the research is quite clear. No, this isn't theoretical research, this is that rarity in economics, proper empirical research. Someone actually going and counting (or, in this case, looking up the numbers others have counted) how many people have phones and what the system used to build phone networks, distribute phones, is. In those Third World countries where there was a State monopoly, or even only one licensed monopolist (sometimes done in the vain hope of keeping the politicians' clammy paws out of the till), the penetration of mobile phones into the population was lower than those places where there were multiple suppliers of handsets and a choice of networks. Yes, markets can be wasteful, in this example the building of several competing networks, with their own towers and repeaters, can be seen, from a certain viewpoint, as being a waste of scarce resources. But if the result is that we get more people with handsets, more people able to communicate, more of those 0.5% boosts to GDP purely as economic efficiency increases, then that's a waste which we're entirely happy to carry: even, a waste which is less than the extra value created and thus not a waste at all.

Which leads us to the conclusion that in order to get that growth we want, the growth that comes from more value being added to resources rather than simply the consumption of more resources, we need to have a market-based system rather than a planned one. To look at an extreme, Paul Krugman again:

But what they actually found was that Soviet growth was based on rapid growth inputs – end of story. The rate of efficiency growth was not only unspectacular, it was well below the rates achieved in Western economies. Indeed, by some estimates, it was virtually nonexistent.

Yes, I know, this isn't what we're usually told, that in order to have green growth, growth that doesn't depend upon more resource-use but growth that comes from more efficient use of resources, we need to have a non-planned economy, we need to have a market economy.

However, we must remember that this desirability of a market-based system is not the same thing as the long hoped for triumph of capitalism over all other socio-economic systems. While the two are often conflated capitalism and markets are very much not the same thing. Capitalism is a method of the ownership of productive assets: markets are a method of exchange.

It is entirely possible for us to have capitalist systems without there being markets. Indeed, this is one of the Marxist defences of communism, that the Soviet Union was not in fact communist it was simply state capitalism and thus cannot be used as the refuting example of why communism doesn't work. Leave aside for a moment the inevitable charge that they're simply whining that they didn't mean it to turn out like that: no, really, we don't think that fairness and justice for all means killing all the smart people and then starving tens of million of others. Look instead at what the Soviet system was really like: there were companies, companies that tried at least to make profits. That the state owned all the companies and all the profits is why we might call it state capitalism but capitalism of a sort it possibly was. There were, however, no markets in this system.

Similarly we could look at what various of the robber barons tried to do in the US in the late 19th century. The setting up of various trusts that would monopolize this or that area of business. Undeniably capitalist, there was most certainly ownership of those means of production but the aim, the intent, was to close down those markets that might, in time, produce competition for said monopolists. Or we could even look at various instances of crony capitalism around the world, where one or another favourite of the powers that be gets a monopoly granted to them. Again, the monopoly's purpose, aim, is to make sure that competition and markets don't break out, thus upsetting the returns to the capitalist. This is how in one African country I can think of it turns out that the President's daughter is the person uniquely qualified to own the mobile telecoms licence and how at one point in the Philippines the coconut industry belonged to one of Marcos' drinking buddies.

We can have capitalism without markets: and clearly we can have agrarian socialism without markets as Pol Pot showed us in Cambodia. Not a notably successful socio-economic system, that one, either.

Running things back the other way, we can certainly have capitalism with markets: that's how we generally view our own, current, societies. And there's no reason why socialism/communism or any other form of collectivism is incompatible with markets. A closer look at our own society, that one we consider capitalist, shows that there's an awful lot of said communalism within it. John Lewis and Waitrose are famously owned by the workers within the business: a workers' cooperative is not exactly a capitalist organization. The Mondragon cooperatives of Spain and the Basque Country show that a similar model can work in heavy industry. The Co-Op itself is owned by the members, another name for the customers. Building Societies by depositors, even law firms are partnerships and thus not capitalist organizations under the classical definition. The great flowering of the British working class was driven by such decidedly non-capitalistic mutually owned organizations which provided everything from burial to unemployment, life and pensions insurance.

We can therefore have socialism and markets as well as capitalism and markets: and looking at the various options, capitalism or not and markets or not, it looks a pretty good bet that it's the market bit that is important, not the capitalism bit. Perhaps, most importantly, that we have a market in methods of organization, that those who wish to work communally are able to, just as those who prefer a bit of oppressing the workers are able to get on with it: and good luck to all who sail in either system and may the best man win.

We should also be careful about bandying about the phrase "free market". While it's a useful and well defined phrase within the confines of academic economics, out here in the real world it leads to some confusion. For of course there is no "free market", there never has been and there never will be. All markets are constrained, all markets are regulated. There are differences in what the constraints are, to be sure, just as there can be differences in who is doing the regulating. We're better off thinking about "freer" markets and less free markets than we are about whether any of them are entirely and absolutely free or not.

Our definitions about being freer also do not, as some might naively think, depend upon how much of the regulating is being done by the State or their appointed agents, the (spit, spit) bureaucrats. What the regulation is, rather than who is doing it, can often be more important.

For example, imagine we had an entirely free market in property. Whatever land or buildings you had could be sold to anyone for them to do as they wished with, you could do as you wished with what is yours and so on. But that, even if it could exist (that is, that you wouldn't immediately lose you property to the bloke with the larger axe, something which has happened in waves as various Romans, Angles, Jutes, Vikings and Normans have turned up on the shores of our green and pleasant land), would not be an unregulated market at all. There would still be regulation, it would just be regulation by some general societal agreement over what is the definition of property? A house? A field? Wood from a forest but not the forest itself? Trout in the river but not drinking water from the same river? The right to graze cattle on the common but not to build upon it? All of those are forms of property rights which have existed, after all. There would also be such a general agreement upon how property is transferred, how we all agree who owns it and so on.

There most certainly have been such privately reached arrangements about property in our past and there are places where much of property law is still of such private provenance. Elinor Ostrom won the Nobel in part for explaining precisely how some of these communal yet non-State systems arise and work. Nor would we really want to describe the work of the Land Registry in keeping (by bureaucrats!) records of who owns what, nor of the courts in enforcing contracts about such ownership, as an impossible State intrusion into free markets: nor would we really describe those two activities as making the market less free.

We would describe, though, the State insisting that you cannot build a house on agricultural land as making the market less free: just as we would describe a communally-derived system which insisted upon the same point as being less free.

Or if we return to the capitalism part and look at corporations: a State which insists that no one but the State can initiate economic activity is certainly at the less than free end of the spectrum. Yet one which insists, via regulation, that there can be no private barriers to the starting of a business, to the initiation of economic activity, we would rightly regard as not making the market less free but more so. Enforcement of monopoly makes markets less free: the eradication of imposed or forced monopoly makes them more free.

It really isn't capitalism that is the crucial part, the thing which allows this continual improvement in productivity and thus gives us economic growth without necessarily having increased resource use. No, it's the experimentation and competition that markets allow which drives this process forward. Nor is it "free markets" that do so, it is properly regulated ones that do. Yes, even to the point that there are useful things that the State and government can do in regulating them to add to what we as producers and consumers will regulate by our own behaviour. That the useful things that can be done by the State are rather fewer in number than those things which said State already attempts to do is an argument in favour of freer markets, certainly, but even here there are exceptions.

Which is the subject of the next and final chapter: what are the regulations, the interventions into markets, which the State must impose if we are to deal with climate change? Not as many as is commonly thought, it is true, but there are some that have to be done for without them we'll never be able to get the job done.

Notes and References
1.   http://web.mit.edu/krugman/www/myth.html

7   Carbon Tax

The science tells us that GHG emissions are an externality;
in other words, our emissions affect the lives of others. When
people do not pay for the consequences of their actions we
have market failure. This is the greatest market failure the
world has seen. It is an externality that goes beyond those of
ordinary congestion or pollution, although many of the same
economic principles apply for its analysis.

Nicholas Stern from the Stern Review.

When markets fail what should we do? Use markets of course.
For the statement is not that markets have failed, that markets
always fail nor even that this is a failure of all possible markets.
It is a statement that this specific market has a failure in it, one
that we'd rather like to correct.

That failure is those externalities, the effects of carbon
emissions upon others, effects which are not included in the prices
we pay in the market economy and therefore are not included in
the incentives which we face to make one or another decision.
As we noted earlier, there's nothing outré or odd about externalities,
nothing new about noting they exist. Further, there's nothing
new or odd about what we do about them – we know what to do
about them. All we've got here is an acknowledgement that here
is a new set of externalities, a new set of reasons why markets
aren't as perfect as we'd like them to be and thus a new set of things
we need to add to such markets so that they are more perfect.

Now it is true that not everyone reads matters in quite this manner. There most certainly are those who insist that as markets have failed here we must abandon markets here. We should turn instead to the wise, near-omniscient and most certainly well-meaning politicians and bureaucrats who rule us and get them to tell us all what to do. To craft regulations, laws, which put us all on the path of righteousness. On the face of it not all that bad an idea until we consider the actual politicians we have: at the time of writing the Coalition hasn't had time to do anything truly stupid so I shall have to stick with examples from the last Labour Government.

John Prescott decided, when crafting his Pathfinder programme, that the best solution to the shortage of reasonably-priced housing in the UK was to demolish some 400,000 reasonably-priced houses. It's very difficult indeed to understand the logic here and such was the now enermined Deputy Prime Minister's rhetorical skills that he never did manage to explain it satisfactorily. "We had to destroy the village in order to save it" is usually used as an example of impenetrable Pentagonese rather than the solution to housing the British working classes.

This disconnect between observable reality and political plans I'm afraid led to my suffering some months of total contempt for Prescott himself. Not, I hasten to add, on the same grounds that Nicholas Soames has tormented him through his parliamentary career, that he was a ship's steward as a young man and that it would thus be terribly funny to shout for a gin and tonic when a fellow MP was speaking in the House. I've served gin and tonics for a living and a fine and honourable job it is: I've even served gin and tonics with one who went on to become one of Soames' fellow Tory MPs although I have a feeling that for Soames the revelation that possible Tory MPs do sometimes work for a living these days would just be evidence that the country really has gone to the dogs.

Such was my contempt that I used my blog to create a little joke upon the intertubes. As and when you use Google to search for something you are offered two choices: either the standard search which brings up those thousands of answers which the search engine is famous for or you can rather short circuit that process and, by clicking the "I'm feeling lucky" button be taken to the one page on the entire internet that Google thinks encapsulates the totality of what you are searching for. The *ne plus ultra*, the epitome, the ultimate example of whatever it is that you have asked about.

There was a time when such results could be manipulated quite easily. I'll not tell you how as it's boring and it doesn't work any more but from my blog, and with the help of a few like-minded souls, it was possible to link a certain word or phrase with a certain page. Thus, the number one result in Google for a certain query, or the result of the "I'm feeling lucky" option, could be fiddled so as to make a joke or even a political point.

Which is how, for a number of months, if you entered the word "fuckwit" into Google, you would be taken to the page of John Prescott, Deputy Prime Minister, at the Number 10 website.

I'm assured that this caused sufficient anger there at the heart of power that an official complaint was made to Google and it's certainly true that the Google algorithm was changed and such jolly japes are no longer possible through the method used. You can find the evidence of this "Googlebomb" in the archives of the internet through Google of course, even if the trick itself no longer works. However, over the years a slightly uncomfortable explanation seems to have become accepted. That this was all something to do with Prescott having come from humble beginnings, or having risen through the trade union movement, even that it was a class attack on such a stout pillar of the working class made politically powerful.

If I could just use this opportunity to set the record straight? This *jeu d'esprit* at the expense of the Baron Prescott of Kingston upon Hull was not motivated by any of those things and as the originator I do know of what I speak. It's just that knocking down cheap houses in order to increase the supply of cheap houses is, well, fuckwitted.

Now that I've that burden of sin off my soul perhaps we should look at environmental regulation? As we've seen earlier in this volume no one actually knows whether recycling saves resources or not because no one has bothered to measure the resources used in recycling. Even if you want to take the absurd position that human labour is not a resource we still don't know whether we are recycling in the most efficient manner, again because no one is measuring the efficiency with which we recycle. So the desirability of regulation to force us to recycle is, at the very best, still unproven.

We've similarly seen that those urging new and interesting forms of energy generation, new forms which will of course be supported again by regulation, bureaucratic dictat, seem confused as to whether jobs are a cost or a benefit: this very strange idea that using more rather than less human labour to achieve a goal being a good thing. And of course these very technologies are enforced by bureaucratic regulation: in the UK through something called the Renewables Obligation. When this is fully up and running in a decade or so it will add £6 billion a year to our collective energy bills: and there's more yet to come with feed-in tariffs, which could add as much again.

Such feed-in tariffs are a perfect example of quite what regulation does bring us. They were announced[1] late in the last, Labour, government's time and they are perverse to the point of idiocy. The aim is to encourage small-scale electricity generation, this stuff that we might be able to generate at home or in a small business, electricity which can then be fed into the grid. For we might not want to use it when we can generate it but others might: we then might want to use some when another is generating. So far so sensible, this is only trade again. But once we've let the government loose on this idea they've ended up giving larger subsidies to the most expensive forms of doing this and small ones to the cheapest. Which really isn't what we want to be doing at all.

We have a number of different technologies available, from rooftop windmills (which never manage to produce as much energy as was used in their manufacture) through water mills, solar cells and so on. We've also our most basic human desire which is to get as much as we can for the least effort: in this case we want the maximum amount of generating capacity for whatever it is that we're willing to spend in subsidy. The subsidy itself is justified by the idea that, well, this is all pretty new stuff so it needs a kick-start. So, being sensible people, what we'd do is set one single feed-in tariff. You generate power and sell it to the grid and you'll get one price. If your system costs more than that subsidy plus the value of the electricity, then you'll lose money and presumably won't do it. If less, then you'll make out like a bandit and thus we get the installation of the most efficient of these various different micro-generation technologies. We've got the most for the least: precisely our aim.

So is that what was done? Those bright people who claim to rule us (you know, the reason that we scour the finest universities in the country and try to get the big brains to join the civil service?) did exactly the opposite. They said that the worse your technology was, the more expensive it was to reduce CO_2 emissions your way, the more money they'd give you.

No, really, this is what they did. They added up the costs of each of the various technologies, worked out how much the electricity would cost if the installer was to get an 8% return on capital and then said that each different technology should get that different price. So if you were to stick in a big hydro system you'd get 4.5p: if you put in a solar PV system you'd get 41.3p.

They've deliberately decided to spend the subsidies in the most inefficient manner possible. Deliberately set out to make little old ladies get the least energy possible for the miserable pittance of their pension they can afford to put into the 'leccie meter.

Sorry, but allow me to run this past you again in a slightly different manner. We've got 45 pence or so that we can spend on subsidizing green energy ... no, just pretend for a moment. There is no such thing as zero emissions energy, although there are a number of different low emission ones. As it happens hydro has about one third the emissions of solar PV over the whole cycle, from digging the holes to get the things to make the systems, to running them, to tearing them down and throwing them away again some decades in the future. For our 45p we'd like as much energy as we can get, please, and by preference, of the lowest possible emissions as well. A sensible system would spend the 45p on ten units of energy from hydro: the system we have will spend 41.3p on one unit of solar energy with a tip left over for the genius who thought this system up (said genius quite possibly being Sir Jonathan Porritt Bt.).

It's not looking all that good for the omniscience of the bureaucrats at the moment is it? Nor their basic comprehension of logic or sums. So perhaps we don't want to go down this route of clever people being stupid in offices* telling us what to do, perhaps we'd like to try working it out for ourselves, using the incentives which a properly designed market will provide?

There's more to it than this as well. As I write, Chris Huhne is telling all and sundry that there will be no state subsidies for nuclear power. Which is a little strange, really, as there are huge subsidies being handed out to all of the other low emissions forms of energy generation. Why not nuclear? Essentially, because an awful lot of people like Huhne don't like nuclear power: the green movement, the whole environmental knitting-yoghurt-flavoured-yurts-from-lentils thing started with opposition to things nuclear, both bombs and power plants. Nuclear was bad before climate change was even a gleam in a campaigner's eye and just because we're about to boil Flipper and drown Bangladesh is no reason to change the mantra that "atoms are bad, M'Kay?". So, while nuclear is indeed low emission, about the same as windmills and hydro, markedly less than solar PV, it simply would not do to allow it to have the same subsidies as the other technologies.

The problem with our "stupidity in offices" made rules, though, is that exactly those rules make it politically possible to do so. The extra costs for the favoured forms of low emission technologies are paid by us directly through our electricity bills. Things like the renewables obligation and the feed-in tariffs. Nuclear isn't allowed access to these subsidies to non-fossil fuel forms of generation. Which in turn means that if there is going to be a subsidy then it needs to come from taxpayers, which means that when Huhne says "no state subsidies" he's, well, he's boxed in nuclear power. Given that just about all of us buy electricity and just about all of us pay taxes it doesn't really matter all that much to us which pocket the subsidy comes from. But by denying nuclear the same subsidy from electricity prices and then insisting that there will be no tax subsidies, what might be the most efficient of the low emission technologies is quite out in the cold.

True, nuclear might not be the most efficient of such technologies but the point here is that we're not even being allowed to find out. For by using the bureaucracy, regulations, to channel our actions we've also allowed ideologues to rig the game towards their preferred solutions: quite possibly to our cost.

If you'd like this point made even more starkly: we've something called the Climate Change Levy. It's a tax on certain forms of electricity generation because they're bad for the climate as they have high CO_2 emissions. As I point out above, nuclear has about the same emissions over the whole cycle as wind or hydro and about one third the emissions of solar PV.

Nuclear pays the Climate Change Levy, solar PV does not. These complicated rules allow those who care enough to understand them to swing decisions to their ideologically favoured position whatever the truth of how well it's all going to solve the problem we actually face.

So, we don't in fact want to use this system to deal with what is, if we accept the IPCC, Stern Review and all that, the greatest problem we currently face as a species or civilization. We would prefer to use some other method, one which has slightly less chance of being entirely cocked up by whichever group of votestealers happen to manage to gain power. Which is why we turn to markets.

That climate change is a market failure does not, as has already been pointed out, mean that all markets have failed, that all possible variants of markets will fail to deal with this problem, only that the market system as currently extant is failing to deal with this specific problem. Our choice of how to use markets to rectify this comes in one of two flavours: we can either shoehorn the problem into current markets or we can create a new market to deal with this problem.

Shoehorning comes from (as Stern goes to great lengths to point out) the acknowledgement that emissions are an externality. They are an effect of our actions which are not currently included in the market prices which guide our actions. As Marshall pointed out at the turn of the last century and as his successor Pigou went on to solve for, we know what to do with these. We add a tax to the action so that market prices now reflect the true costs of said actions. We've even got a number from Stern as to what that tax should be: $80 per tonne CO_2.

We ought to take a little detour here to discuss the validity of that $80 and, truth be told, there's not a great deal of validity to it. The Stern Review plays a number of tricks to get to it. The first and most obvious is that all of the calculations are based upon only one of the four families of possible economic (and thus emissions scenarios) that the IPCC itself considers. You don't have to be as cynical as I am (although I prefer "realist" when trying to describe my bleak and total cynicism about the actions of politicians and their hirelings) to guess that he used the very worst of those economic models, the family that produces hugely high emissions by comparison with the others.

No, sorry, let me backtrack a little. He does use another set of emissions: one he made up for the task. One that the IPCC hasn't considered and one which is, yes, you guessed it (see, told you, realism) even worse. So our $80 is based upon as bad as the IPCC thinks it could be and worse: no consideration is given to the idea that it might not be that bad but that's still part of where we get our $80 from.

The second trick is that Stern essentially invents a new way of dealing with discount rates. No, we'll not go there, it's very long and boring but let's just say that his treatment of this issue received a great deal of commentary ("commentary" is the polite way economists describe making the point "You did what? But, but, don't you understand the implications of that? Buffoon!**") from economists who had actually been working within the IPCC structure, economists like Sir Partha Dasgupta and Richard Tol.

A third trick, well, no, not really a trick, rather a gross oversight, comes in the treatment of the technological and capital cycles.

One of the great arguments in economics (it's at the heart of what all those talking heads on the TV screens are shouting about, recession, unemployment, government spending and the rest) is about how quickly things happen. If this bit of the economy over here changes then how long does it take for that bit over there to adjust to it? A Keynesian (or even a New Keynesian, although for slightly different reasons) will think that in a recession then wages won't change, won't change quickly enough at least, which is why we get unemployment. A New

Classical (or again, for slightly different reasons, a Real Business Cycle theorist) would say that of course wages will adjust, near instantaneously and thus there cannot be recessions and whatever it is that we're seeing is caused by something else. Entirely. No doubt at all. That however is macroeconomics, as PJ O'Rourke pointed out, the part of the subject where we're reasonably certain that we don't know what we're talking about.

However, microeconomics (the bit where we know a bit is correct at least) covers the same point about how quickly things happen. For example, we know that the short and long term effects of tax changes are different: it takes time for people to change their behaviour. We also know that we've something we can call the technological cycle: how long does it take to get some new whizzy way of doing something into the hands of people who will use it to do whizzy things? Specifically, here, with climate change, we'd like to know how long it takes to get some nice new low-carbon technology thought about, developed, tested, manufactured and thus really ready for use. Given that windmills have been around in Europe since at least the 12th century we can see that it can be a fairly considerable amount of time.

As a matter of personal experience, while perhaps not eight centuries and counting, it's still rather longer than many seem to think. My own day job involves wholesaling the metal scandium (no, don't worry, no one else has ever heard of it either) and there's good reason to think that scandium oxide will be the "miracle ingredient" which makes a certain type of fuel cell, solid oxide fuel cells (SOFC), a commercial reality. They're not a new method of generating power, for a start they're not new, they were first devised in the 1840s. They're also not likely to be all that good at generating power either, for you either run them off various flavours of fossil fuels (natural gas being a good one) or hydrogen. And hydrogen currently either comes from natural gas or you've got to electrolyse water to get it which in itself requires more energy than you're going to get from your fuel cell. However, what such fuel cells might be very good at indeed is as part of the storage system which any largely renewables-based power generation system needs.

Our problem with windmills is that the wind doesn't always blow (for the UK it's actually been shown that cold winter days, peak energy demand time, are correlated with high pressure areas over the country meaning little to no wind. One paper went so far as to point out that on one specific such cold winter's day there wasn't a single windmill in the country which could produce any electricity at all), and for solar that the sun doesn't always shine (it's not really necessary for me to point this out to inhabitants of the UK is it?). And winds blow and the sun shines quite wonderfully at other times, unfortunately often at times when we don't really want to use any power. Given that we don't use air conditioning (preferring woolly vests and pith helmets as ways of dealing with the summer heat) those halcyon days of sunshine with a light breeze we get once or twice every 36 months mean that peak energy generation possibilities coincide with the lowest energy consumption of the entire year. What we really want therefore is some system of being able to generate when we can and consume when we wish to, something which requires a storage system.

Storage systems for electricity are, on a small scale, no problem at all. We've all sorts of methods of that, capacitors, batteries and the like. We've also a few ideas about how to do it on a large scale, pumping water uphill for example, letting it run down through a turbine when we want it. One other contender for this job is these very SOFCs. When we've got electricity being generated let's use it to crack water into hydrogen and oxygen and store the hydrogen. Then, when we want to do something, run the hydrogen through the fuel cell and we have electricity. You can go and buy systems to do this today; they are, however, quite extraordinarily expensive as yet.

That scandium was a good idea in such systems was realized by Westinghouse in the late 1980s. They even went so far as to investigate new methods of scandium extraction, patenting a couple of such methods in 1991, in order to provide themselves with the supply they thought they would need. Sadly, their approach didn't really work out and the idea was dropped (their basic idea was fine, it's just that the technology to do it efficiently didn't exist then but arguably does now as others are experimenting with). Others have looked at such SOFCs as a replacement for the petrol engine but that rather stuttered when the ceramic plates that make up the cell kept cracking as they moved from off (cold) to on (hot). That particular problem was solved at St Andrew's University in Scotland in 2002. The first real commercial manufacturer, Bloom Energy, of such SOFCs only announced its product earlier this year, the Bloom Box, after 7 years of development and $400 million of venture capital funding. This still isn't quite right either but it's getting there: more work needs to be done on the manufacturing process to make it cheap rather than just effective, for example. They've also got those same cracking problems but as the St Andrew's solution would require them to use less rather than more of my beloved scandium I won't tell them if you don't.

Two things to take from this over detailed look at one renewables technology: the first is that before we'd even had the Rio conference that kicked off the whole global warming concern (back when it still was called global warming rather than climate change for example), before Kyoto, the R&D departments of the world were already working on solutions. It isn't actually necessary for there to be a tax now, an insistence now, for people to try and create the technologies which will solve such problems. Even just a concern (not even a certain knowledge) that such taxes and insistences will happen in the future are enough to start people pondering on how they might make some cash out of it all.

The second is of course that these things take time. Leaving aside the 1840s part, we're still 20 years into the modern era of developing solid oxide fuel cells and we're (in my estimation at least, as a semi-insider in the business) another five, maybe ten, years away from having a truly mass market and affordable product. Screaming that we must do something now doesn't change this fact that developing new technologies simply takes time.

The second oversight of Stern's comes from the capital cycle. We're really not talking about doom and destruction (whatever the hysterics say) if we don't stop burning coal next week: nor even next year nor next decade. It might well be true that we're going to have some problems if we're still burning coal in 80 years' time however. But no coal-fired power station we have today is going to be operating in 80 years' time: the capital cycle means that we will have torn it down and built another one. What we would like therefore is not a policy (or tax, or permits, or regulations) that stop us using the coal plants we've already spent billions building, but rather some means of making sure that when we get around to replacing what we've already paid for we don't build a coal plant but some lovely cuddly non fossil fuel plant instead.

Mix these two oversights together and rather than what Stern suggested, that $80 tax right now, what we might prefer is some credible commitment to there being a tax in the future, which will both stimulate research now and make sure that replacement as the capital cycle turns is with non-fossil plants. That is, we might prefer the suggestions of William Nordhaus (an economist who is actually an expert in matters climate change rather than an international bureaucrat plucked to write a report to order on the subject), that we have a low-carbon tax now of say $5 per tonne but which rises over the decades to perhaps $250 per tonne in the 2040s. For what is the point of taxing us all now if it simply takes time for everyone to change their behaviour? To research and manufacture the new technologies, for us to wear out the plants we have already paid for?

Having said why we probably shouldn't like Stern's calculations of how much we should all be paying for emissions, let's go on and simply accept that he's right. Yes, OK, it should be $80 a tonne, right now, as a tax on them. That's how we're going to deal with climate change. So what, actually, is it that we need to do now?

Strangely, not a lot really. For example, we've Air Passenger Duty which is there to make up for the fact that there's no tax on aviation fuel. The APD we have to pay on every ticket in and out of the UK is pretty much (the "pretty much" rather than exactly is because it's charged in bands and thus isn't absolutely exact for each and every flight) the emissions that our flight creates times that $80 per tonne. So, we're done as far as flying is concerned. We don't have to have restrictive laws on the number of airports or flights, Plane Stupid are being plain stupid for their concern has already been addressed: everyone is now paying the true cost of their trips to get drunk in countries with less than Puritan levels of alcohol taxation. We have, by getting the costs of emissions into market prices, managed to get to the socially optimal amount of emissions.

Ah, no, I can see that I need to explain a little more here. We need, sadly, to look a little more at the economics of Pigou Taxes, those things which started out with the effect of one farmer's rabbits on the fields of another.

There are those out there who think that green taxation means we should just raise taxes until people stop doing whatever it is that isn't green. If, say, cars damage the environment then taxation on cars should be raised to the point that everyone uses the bus or a train. But this isn't actually what we're trying to achieve: we're not looking to have no pollution, no emissions, at all. We're looking to have the optimal amount of emissions, the socially optimal amount. That's what all of the work that went into the Stern Review is all about.

We recognize that yes, there are costs to emissions and other pollutions. But we also recognize that there are benefits to being able to so emit or pollute. What the end result of Stern gives us, that $80 figure, is a way of finding the optimal amount of emissions. Yes, inundating Pacific islands in a century's time imposes a cost upon others through our actions today. But not inundating Pacific islands in a century's time imposes costs upon us today. We have to travel with the oiks on the railways or fight our way through the crowds of crumblies wielding their free OAP bus passes for example. It's true that these costs to us are not very high: but then neither are the costs of our taking the car all that high to those in the future. In aggregate, yes, the damage is large but then also in aggregate, all of us piling onto the bus or train is also a high cost (quite apart from the fact that there isn't enough public transport to carry us all, meaning that for some the cost is no travel at all). But when we've global emissions of 5 billion tonnes a year of CO_2 the couple of kilos from driving across town really isn't all that much of a burden, just as the having to scrape lager fuelled vomit off the train seat isn't, in the greater scale of things, all that much of a problem.

So, what we really want to do is have people not stop anything and everything which causes emissions: we want them to stop doing those things where the emissions have a higher cost than the benefits which are gained from them. Similarly, we do want people to do the things where the benefits are higher than the costs. This is back to our old friend, the cost benefit analysis again. Do those things where the benefits are higher than the costs, do not do those things where the costs are higher than the benefits. In this manner we are able to maximize human utility, human happiness, the joy and gloriousness of the world as experienced.

At one extreme we might say that my driving to get the bread for lunch isn't worth the emissions, the damage done to others in the future from them. OK, so I'll cycle, as I do (and ignore that pesky guy who keeps insisting that walking and cycling use more energy than driving because humans are very inefficient processors of energy, unlike cars). But what about the ambulance on a run to take the woman with pre-eclampsia for treatment? Same litre of petrol being used, same couple of kilos of emissions, the same damage done to those in the future: but two lives saved in the here and now. We would like that ambulance run to be done, even at the cost to those in the future. Yes, even at that cost to the rain forests, biodiversity, the glaciers and Gaia herself.

This is what the Pigou Tax does. We've introduced into the price system, by adding that $80 to any and everything which produces a tonne of CO_2, the effects of our actions on that future. We will therefore only take such actions where the benefit to ourselves outweighs those costs. We are therefore only going to undertake actions which increase human utility over time, and we'll have been dissuaded from doing the ones where the costs are higher than the benefits by having to pay those costs. So slap on the carbon tax and we're done.

To which there are two interesting little additions. You might argue, as many do, that $80 isn't the right price. The forests are worth more than that alone, the lemurs of Madagascar, that island off Kiribati, their value is much higher than Stern or anyone else has realized and that thus the tax should be much higher. If that is the way you think then by all means you should indeed be arguing that: but do understand that we're not using your valuation here. You've got one seven billionth of the vote, just like everyone else. For, despite what everyone from St Thomas Aquinas to Karl Marx said, there is no such thing as absolute value, no true value that can be divined. There is simply the value that human beings place upon things:

as it is we humans doing the valuing, the value can only be the value that we apply. What was measured by Stern and what will be measured by any other attempt is the entirely subjective value that currently extant humans place upon whatever it is being valued. Which means that all you have to do to get that correct tax level higher is go off and convince the other 6,999,999,999 of us that your value is correct and when you have so done, it will be. Good luck.

The second is that we've now got a number for what is the optimal tax level. No, not the one that the government wants in order to pay for their schemes, not one that they can get away with, but one that is "correct" in a larger sense. One that correctly balances present and future. Which, when we look at petrol prices, is really rather fun.

Back when Ken Clarke was last in Government (and how long ago that both seems and is) he brought in the fuel duty escalator. This was expressly, in his words "to meet our Rio commitments". That is, it's a Pigou Tax upon the emissions from petrol. We can, using our $80 figure, work out pretty quickly what the correct level of tax should be for the emissions from the burning of a litre of petrol: around 11p. We can also look at how much extra duty has been added by the fuel duty escalator over the years: around 23p. Thus, to get to our correct level of taxation upon petrol we should cut fuel duty by 12p a litre: very different indeed from what we keep being told, isn't it?

Yes, there are other things that fuel duty pays for, the roads, congestion, particularate pollution, but these are already paid for in the other 30p of fuel duty we pay on each litre. We can test that we're about right here: Gordon Brown brought in a special duty class for a time for biodiesel. Such a fuel would have all of the other problems and costs but not the CO_2 emissions (because the plants would absorb the gas while growing). The tax break was 20p a litre. One of the interesting outcomes therefore of this whole investigation into the correct level of environmental taxes is that petrol duty should be cut, not raised.

Note also that it doesn't matter what such taxes are spent upon: we are not trying to compensate those polluted, pay to clean up the pollution nor even stop people polluting at all. We are trying, solely and only, to make sure that because of the prices they pay people are only undertaking actions where the benefits are greater than the costs. This is achieved by exactly the tax and is thus all that we need to do.

If saving the planet by lowering taxes sounds just too suspiciously neo-liberal for you then I should go on to point out that while the general tax level is indeed about correct, it's not quite levied on all of the right things. It does depend a little upon who you talk to and quite how they're running the numbers (should we include the emissions in imports? We are "consuming" that CO_2 after all. But if that, then should we exclude the emissions from exports?) but a rough and ready guide to UK emissions is 500 million tonnes a year. At our $80, depending upon the exchange rate, that's about £25 billion a year that we should be paying in emissions related Pigou Taxes. As it happens, that's not all that far away from how much we are paying in emissions taxes. Again, it depends on exactly how you run the numbers; certainly we should include Landfill Tax, the Climate Change Levy, that extra fuel duty, Air Passenger Duty. But also perhaps some or all of the costs of the Renewables Obligation, the coming costs of the cap-and-trade system (which we shall come to) and innumerable smaller impositions upon us. Certainly we're in the right order of magnitude, we're paying at least ten billion pounds and there are enthusiasts (like the Taxpayers' Alliance) who claim we are paying nearer £30 billion. Though farmers are almost certainly paying too little (both cow farts and nitrogen from fertilizers are significant emissions sources) car drivers seem to be paying too much even if the overall level is about right.

There is also another way entirely of doing this, rather than shoehorning concern and prices for emissions into current market prices, why not create an entirely new market for emissions? This is what is known as cap-and-trade and is something which is being done at a European, rather than national, level. It is, of course, as a European Union project even more desperately screwed up than anything that our own homegrown politicians have managed.

In theory, the effects of a properly designed cap-and-trade system should be exactly the same as a properly designed carbon tax system. We should, for the same costs, get the same reductions in emissions. However, cap-and-trade systems allow politicians more input into how the system is designed, meaning that of course it won't be properly designed. One of the most obvious points has been made by Greg Mankiw, an economics professor at Harvard (and formerly Chairman of the Council of Economic Advisors for George Bush II... just to show that this isn't a left/right thing, rather an economists v. everyone else kind of thing) which is that for a cap-and-trade system to work all of the permits must be auctioned off. If we just give them away then the effect of cap-and-trade is like that of a carbon tax plus lots of corporate pork. Guess what's happening in the EU system? Yes, most of the permits are being given away so we've got lots and lots of lovely favours that politicians can do for favoured groups. How unlike any political system that we'd actually desire.

There is one major difference between the two methods though, of sticking emission constrictions into the current market through a tax or creating a new market by cap-and-trade. In cap-and-trade we say that we're only going to permit so many emissions: anyone who wants to emit must have a permit to do so. They have to go out into the market (if the politicians didn't give them away) and buy such a permit. So, we know what the emissions will be, because we've limited them. But we don't know what the cost will be because we just don't know how much people will bid for those permits.

With the carbon tax we do know what the cost will be: it'll be the tax. But we don't know how effective that will be in reducing emissions.

It's absolutely possible to make good arguments on either side of this: which should we prefer, tax or cap-and-trade? I personally come down on the side of tax for three reasons, the first being that you might already have noted my distaste for politicians. A system that keeps them as far out of the way as possible meets my basic test of being more likely to work. The fewer policy wonks we have trying to perfect something the less likely it is that they'll screw it up.

The second is that we don't really know exactly what the cost of carbon emissions is. Not just that perhaps the Stern Review is wrong but that no number we come up with for something as chaotic, non-linear and complex as the economy is ever actually going to be right. We might get it about right but that's the very best we will be able to do. FA Hayek has had some things to say about this at times, the impossibility of our ever collecting enough information and then processing it in time to be able to do anything useful about planning the economy in any sort of detail. This would seem to militate in favour of a cap-and-trade system therefore: but the climate is just as chaotic, non-linear and complex as the economy is. So we also don't know what the correct level of emissions is, cannot, in the same way that we can't know the correct price. But we've been looking at the economy for a lot longer than we have the climate, we know more about the various interactions, there are more people doing the looking and in general the science is more developed. It is impossible for us to know either, economy or climate, price or emissions levels, correctly and in detail, but given the current state of the respective disciplines we're more likely to get the price right.

The third is the clinching reason for me and it's one that can make heads explode when spelt out. What we're trying to do is to maximize human utility, as above, maximise that amount of happiness that it's possible for us all, both now and in the future, to have. We've already sold the pass that we're not going to have no climate change at all: we're trying to get to the right amount of climate change, the one that allows us to do the things which benefit us more than they will harm those in the future. This means that it isn't actually the level of emissions, nor the level of CO_2 in the atmosphere (either the flow, the increase, or the stock of past flows and increases) that we're trying to target. We're trying to target the cost of how we deal with climate change. No, really, we are, we really are saying that it's possible that reducing our emissions to some level will be too expensive: some people in the future are just going to have to put up with a changed climate and adapt to it.

This point is easily proven: we aren't saying that we're going to stop emissions today. We are saying that we'll try and reduce them but doing it all today is just too expensive, thus some of you in the future will have to adapt while we work out how to reduce emissions. The determinant of whether it's too expensive is whether the costs of reducing emissions are greater than the future benefits of doing so. If it costs us $100 to reduce a tonne of emissions now but the cost in the future is only $80 then we don't want to do it. Humanity, over time, has just been made $20 poorer by our reducing those emissions. Conversely, if it costs us $60 to reduce then humanity will be made $20 richer. And it is the tax method which puts that into our decision-making process, gives us the incentives to change our behaviour to properly balance present and future. Thus we should be using the tax system, not cap-and-trade.

Just to digress, with the cap-and-trade system that is in place, that European Union Trading system, we're seeing complaints that the price of the permits is too low. It can cost as little as 10 euros to buy a permit to emit while we've got our known cost of $80. Thus we are being told by various types (oddly, rather a Baptist and Bootlegger mixture of greenie organizations and a couple of the power companies) that there should be a minimum price imposed for these permits: 30 euro is often talked about. Why the power companies are saying this is clear: they're getting these permits largely for free and if there's a minimum price then they get to make pots of money. The environmentalists seem to be getting tripped up by their ignorance of economics, though. They say that we need higher permit prices in order to get people to change their behaviour, to curb emissions. But that's not how a cap-and-trade system works at all. We've the cap, that's what reduces emissions. The price of the permit is telling us what is the price of reducing those emissions. Thus, far from a high permit price being a good thing, it's terrible. Yes, of course we want to reduce emissions but we want to do it at the lowest price possible. We'd prefer the permit price to be 10 euro rather than 30 because this shows that reducing emissions to the level of the cap is cheaper than we had thought it was going to be. Heck, we'd prefer a price of 1 cent per permit than one of 2 cents per permit.

That's probably enough nitpicking even for a book on such points: the real takeaway line from all of this is that whether we use a cap-and-trade system or a carbon tax, that's all we need to do. Yes, we need to make sure that all of the different emission sources are covered by one or the other, we need to make sure that we've got our tax burden correctly distributed (not too much on petrol and too little on farming for example), but once we've done that, we're done. We just need to wait and see how those new incentives cascade through the economy.

If we use cap-and-trade then as long as we've set the cap at the correct amount (and if we cannot do that then politics isn't all that much use to use in solving this problem) then, well, we've capped emissions at the correct amount. We don't want to start mithering about whether this sector or that sector needs to reduce its emission more than it is: nor about whether one or another sector is still increasing emissions. We do see this with respect to aviation emissions, lurid stories about how if everything keeps going like it is and we manage to cut total emissions by 80% by 2050 then the only emissions we'll be making will be from aviation.

But what's wrong with that? As long as we're below the total cap then we want to use those emissions we are allowed to make on the things that we value most. No, not what someone else values and most certainly not on what someone else thinks we should value: but on what we ourselves value. Imagine that we are indeed all using renewable energy, cars all run on electricity or fuel cells, houses are perfectly insulated and cows no longer burp. Or even that making those things so costs us some of our income, costs which we are happy to bear as that means that we can indeed use that 20% of current emissions, that 100 million tonnes a year, to fly off to see the world. We're getting the greatest value we can from that scarce resource: isn't that what the whole game is about? Getting the most from what we have available?

If we go the carbon tax route then by pricing in the emissions into our daily market transactions we've made sure that only those emissions which create more benefit than they do harm will happen. We don't actually want to do any more than that: we are, again, trying to maximize human happiness, human utility, by making the most of what we've got. Getting the tax level right means that we've done that over the generations as well as only just now. We don't want to spend $1,070 preventing a tonne of CO_2 being emitted (which is, would you believe, the German cost of their feed-in tariffs for solar power subsidies) when the damage it will do is only $80. We do want to stop someone getting $10 of benefit by dumping $80 of costs on the future and the tax does that very nicely.

As I say though, once we've done either or both of these things that really is all we need to do. And if you don't believe me, perhaps you'll believe someone from the IPCC? Richard Tol has been, over the four IPCC reports, variously a ghost writer, author, lead author, senior lead author and co-ordinating author: that's all ranks and he's worked on all sectors over the different reports. He's also an economist and his work has been, not unnaturally, on the economics of what is going on. Here's the crunch lines from a recent paper of his:[2]

In sum, these results call for a measured policy of greenhouse gas emission reduction. There is reason to believe that European climate policy is overly ambitious. Climate policy outside Europe is surely not ambitious enough.

There's still a way to go for the Americans, Indians, Chinese and all the rest but for us here in Europe and even more so for us here in the UK, we're done. We've done what needed to be done, we've put the price of emissions into the market prices we face each day and that is, really, all we've ever needed to do. We will now have the socially optimal amount of climate change which is the very best that anyone can possibly ask of us. Possibly, even, if Tol is correct, we'll have not enough climate change.

Notes and References

1. http://www.decc.gov.uk/assets/decc/what%20we%20do/uk%20 energy%20supply/energy%20mix/renewable%20energy/policy/ fits/1_20100304142317_e_@@_tableoftariffsupto2013.pdf
2. http://www.voxeu.org/index.php?q=node/4245
   *   This is "Worstall's definition of bureaucracy: clever people being stupid in offices".
   **   Buffoon is not a word, or even a thought, which would come from one such as Sir Partha Dasgupta and I am most certainly not implying that he said or thought of such a thing in relation to the then Sir Nicholas Stern. Even though he did disagree, at length, with his treatment of discount rates. Richard Tol however . . .

Index